WEDDED TO TROUBLE

BOOKS BY SARAH OSBORNE

The Ditie Brown Mysteries
Too Many Crooks Spoil the Plot
Into the Frying Pan
Murder Most Southern
The House of Good and Evil
Wedded to Trouble

Flo and Maude Christmas Cozies
Flo and Maude Christmas Capers
Flo and Maude Save a Santa

WEDDED TO TROUBLE

A DITIE BROWN MYSTERY, BOOK 5

SARAH OSBORNE

First Edition: 2022

ISBN:978-1-7375565-7-2 (Ebook edition)

ISBN:978-1-7375565-8-9 (Paperback edition)

Imprint: Independently published

Printed in the United States of America

Created with Vellum

For Laurie Pocius who gave me the wonderful title for this book and continues to be one of my most conscientious and enthusiastic beta readers

ACKNOWLEDGMENTS

As always, my first thank you goes to my beta readers who continue to amaze me with their willingness to read my writing in progress and give me candid comments on what works and what needs to be improved. For this book they include the following: Sally Byrne, Sue Eliason, Kathy McCarthy, Kathy Mosesian, Laurie Pocius, Lynne Rozsa, Jean Wentzell.

Once again, my cover designer, Aisha Akeju at https://aisha-o.com/book-covers has created a beautiful cover that reflects my story. Many thanks to her.

The bouquet on the cover is from an actual wedding I attended and was photographed by Hannah Berglund. Check her out at www.hannahberglund.photography if you need outstanding photos for some important event.

My 241 Fitness buddies and Wendy Bryant continue to support my efforts and do their best to help me stay fit. I couldn't do it without them.

And last but never least, my love and thanks to Dan and Alix for their ongoing encouragement.

MEMORABLE LOVE AND MARRIAGE QUOTES

"*Love doesn't just sit there, like a stone, it has to be made, like bread; remade all the time, made new.*"

– Ursula K. Le Guin, *The Lathe of Heaven*

"*You know you're in love when you can't fall asleep because reality is finally better than your dreams.*"

– Dr. Seuss

"*Let there be spaces in your togetherness, And let the winds of the heavens dance between you. Love one another but make not a bond of love: Let it be rather a moving sea between the shores of your souls.*"

– Kahlil Gibran, *The Prophet*

"*To be fully seen by somebody, then, and be loved anyhow–this is a human offering that can border on miraculous.*"

– Elizabeth Gilbert, *Committed: A Skeptic Makes Peace with Marriage*

1

Mason held my hand for a long time after we finished packing the car. When he finally let go, both our hands were sticky from the sweltering heat that was Atlanta in the summer.

"It's all perfect," I said, and he grabbed me in a long embrace that made both of our bodies tingle with perspiration. And something more.

The Atlanta air was hot and humid, the way it always was in July—even at eight in the morning. Mason stood back, staring first at me and then at the street where I lived. In a matter of days it would be the place he lived as well, a quiet street in Virginia-Highland, ten minutes from downtown Atlanta. We'd had a discussion about whether or not we'd move to a new house once we were married, but neither of us wanted that. The children were settled in the house, and my best friend, Lurleen, was minutes away. It was a no-brainer, really. Mason would sell his condo in downtown Atlanta and move in with the kids and animals waiting for him to become a permanent member of our household.

"Can you believe it, Ditie?" he asked. He rubbed a hand

back and forth over his bald head—the way he did when he was excited or anxious about something.

"No, I can't," I said. "It's finally happening. Are you nervous, having second thoughts? Trying to make some hair grow?"

Mason laughed. "You take me as I am—bald, in need of reading glasses, not as fit as I was ten years ago—that's one of a hundred things I love about you."

"Mason, you do the same for me—you relish me the way God made me. Not every man would love my round figure, my curly hair that has a mind of its own, my outspoken nature. I am so lucky!"

"*We* are so lucky, and in two days, we'll be married."

"'Yes. I'll be in my favorite vacation spot with my favorite people, marrying the man of my dreams."

For years, I'd been going to Cape Azul, a spit of land on the Gulf of Mexico between Apalachicola and Panama City. I'd found the tiny peninsula one summer when I stayed on St. George Island and went exploring along the Florida Panhandle. Once I discovered Cape Azul, I decided this was where I would come every summer.

I'd kept that promise, and every year I spent a week on the undiscovered twenty-mile stretch of land that ended in a state park. No high rises. No big hotels. Just houses and rental units and sandy beach that seemed to go on forever.

Every year Cape Azul changed. Sometimes the beach eroded from heavy storms, sometimes more sand was brought in, but I was never disappointed by what I found there. It was always different and always the same. And it always renewed me—the warm water of the Gulf, the pre-dawn darkness that suddenly burst into day, the place where time stood still and there was no need to rush to do anything.

I suppose the real miracle was that Mason and the kids loved it as much as I did.

This summer, our closest friends, Lurleen and Danny, had

gone down two days early with the kids to get everything ready for our wedding. I couldn't wait to join them.

Lucie and Jason were officially my children now and soon they'd be Mason's—that was another miracle. The adoption had gone without a glitch once all concerned parties had renounced their rights and given us their blessing. Even Jason's biological father, Charlie Flack, was behaving himself.

He'd sent Jason a beautiful adoption card from California—who knew such things existed? In it he said how pleased he was that I was adopting Jason and that soon Jason would have two fathers to look after him. Charlie promised visits in the future, but we'd have to see if he could keep those promises.

The adoption ceremony itself was a small affair in a judge's chamber. Mason, Lurleen, Danny and I stood in a semi-circle around Lucie and Jason as the judge announced the children were now *my* children, officially and forever. Even thinking about that ceremony three months later made me cry.

Mason handed me a handkerchief, always the gentleman and always with what seemed to be an endless supply of cotton handkerchiefs. "Why are you crying?" he asked.

"I'm so happy," I said. "The kids, you, everything."

"Me too," Mason said. He smiled that smile of his that made his eyes crinkle. His bald head and that smile made him look like a young Dalai Lama—full of mischief, joy and good feeling. "I can't wait to be married to you," he said.

I squeezed his hand. "Then let's get going. It's a seven hour drive from Atlanta to Cape Azul. The Gulf and the kids are waiting for us."

"You know it will be closer to six if I'm driving," he said.

The one argument Mason and I had when we headed for Cape Azul was which route to take. I liked the scenic route that took us through Tallahassee, then Apalachicola and allowed me to see the ocean hours before we made it to our rented condo. Mason liked the inland route—faster. Today I let him win that argument because I wanted to get there as soon as possible. I wanted plenty of time to prepare for our wedding.

My dog, Hermione, knew something was up—she saw the bags go in the trunk of my old Toyota. She whined a little each time we left the house with boxes of food and supplies, and she tried to sit on our laps when we took a break on the couch. She was too big to manage that and had to settle for a cuddle next to us. We rubbed her golden fur and told her she was the best dog ever, which she was.

Luckily, she loved our pet sitter—the daughter of our next door neighbor. She was home from college and eager to have her own space for a couple of weeks. My big yellow cat, Majestic, stayed out of sight until the last minute. He was making his usual point that no one should disrupt his life in any way. I scooped him up before we left and asked him to be a good boy. He stood at the screen door beside Hermione as we climbed in the car and took off.

We stopped at Krispy Kreme for coffee and doughnuts before we left Atlanta. I got my usual, a chocolate covered cake and a lemon filled. Mason settled on the glazed doughnuts that were coming off the conveyor belt inside the store.

"Can't get fresher than these," he said as he sat in the driver's seat of my car and devoured his first one. "Of course they don't compare with yours, Ditie."

"Nonsense. These are the best doughnuts in the world. It's a good thing we don't live any closer to Krispy Kreme than we do. And, Mason, it's time we have a small heart-to-heart chat."

"Uh oh," Mason said, wiping his face with a napkin. "What have I done?"

"Nothing," I said, "but you should know me well enough to realize I don't need empty compliments about my food or my baking. If I make something delicious, I'll be the first to say that, and if someone else makes something better than I do—like these doughnuts—I'm happy to acknowledge that."

Mason smiled. "I do know you well enough for that—my always-honest Ditie."

I sat back with my cup of coffee and my doughnuts as Mason got us out of Atlanta traffic. I enjoyed the pleasure I felt every summer when I left sultry Atlanta and headed for the Gulf of Mexico. On top of that was the joy of being almost married to the man I loved.

We'd been very fortunate when we looked into places to rent on Cape Azul—ones that could accommodate us and our friends for the wedding. We found two houses side by side that would be perfect. Both were right on the beach. The one Mason and I would stay in was smaller, but it had a lower and upper deck, so we could see the ocean from upstairs and down. The other house had ten bedrooms and a sweeping veranda. The ten bedrooms would more than meet the needs of our small cluster of family and friends. The veranda would be where we'd have the wedding if a sudden storm popped up and we couldn't have the ceremony on the beach.

I wanted to be married at first light, my special time of day, but most of our guests weren't early risers. We settled on late morning for the big event. We'd follow that with a wedding brunch prepared by Anna Hayes, the wonderful chef who was now the manager of Lurleen's cooking school.

The invitation list was small—fewer than twenty guests in total. When we returned to Atlanta, we'd have a bigger party—an open house.

Amazingly, my minister from the Unitarian Church decided he needed a short summer vacation and would be happy to marry us. He and his wife rented a place at Mexico Beach, half an hour away, and he agreed to come early on the wedding morning to perform the ceremony.

The rest of the party included my brother, Tommy, and his partner, Josh, plus my boss from the refugee clinic, Vic, and her husband. Her kids were now in college and had other things to do with their summer.

Mason's mother, Eddy, was coming with Mason's two sons, their wives, and their two young children. We hadn't seen much of either son or their families. The older boy, Kyle, was in the diplomatic service and only recently returned from Thailand. The younger son, Brian, was a high school teacher and coach living in Texas.

Lucie was excited about babysitting the little ones who were both under three. She was hoping to convince me that, as an almost-twelve year old, she was ready to set up her own babysitting business. At eight, Jason was happy to be a boy running free. I allowed each of them to bring their best friends —Hannah and Noah.

Lurleen insisted I do no cooking or baking for the wedding —Anna would handle it. Anna had written me a long letter about how excited she was to come to the wedding, to bake the wedding cake and cook all the meals. She said she could never repay us for what Lurleen and I had done for her. What we'd done for her was offer her a cooking school in Atlanta, something her former boss in South Carolina had promised but never delivered.

"You're quiet," Mason said after an hour of driving, which got us well on our way to Columbus, Georgia and the Alabama border.

"I was thinking about our guests," I said. "I hope they'll be comfortable."

"They're family," Mason said. "Even the friends are more like family. They'll be fine, more than fine if Lurleen is directing things. They'll be treated like royalty."

I smiled at that. Lurleen would be a gracious and gorgeous host, and I had no doubt Anna would be fixing delicious food for those who came a day early.

"Is there anything we need to sort through on the way down?" I asked.

"Are you asking if I've memorized my vows or if I have the ring? Yes to both those questions. What about you?"

"I'm good. I wish my parents were alive to see how happy we are. My dad especially. Mom might have needed to critique everything and find fault, but Dad would have been so pleased about all of it. I hope he's out there somewhere nodding his approval."

I sighed and settled back in my seat. Mason and I both had two weeks off, and I'd never before taken a two-week vacation from the refugee clinic. Vic said she had some part-time pediatricians coming on board and this would be a good way for them to get their feet wet.

Mason and I had considered taking an exotic honeymoon somewhere but neither of us wanted that. Lurleen came up with the perfect solution. She and Danny would look after the kids and entertain them for two weeks. They'd move down the beach after the wedding to the rental unit we usually booked when we came each summer. That way we could have our privacy and still see them as often as we liked.

Danny and Lurleen were the perfect godparents—they'd let the kids get away with far more than we would, but they'd keep them safe.

"I have nothing to worry about," I said and kissed Mason on the cheek.

My cell went off at that moment. It was Lurleen.

"I hate to bother you on your way down, but we have a slight problem here."

2

A slight problem? Lurleen's slight problem could be anything from flowers that didn't arrive to a murder we needed to investigate. I held my breath.

"It's nothing really," she said, "but I thought you should be informed. I'm a little worried about our landlord—the owner of the houses you rented. He stopped by to say there must have been some misunderstanding. He has a new renter who wants the big house tomorrow. I told him that was impossible and that we had the house for another three days, so people wouldn't feel they needed to rush off. I showed him the rental agreement. He was a smallish man but with a very big mouth. He huffed and puffed, but when I asked Danny to stand beside me, he left."

I wasn't surprised by that. Danny was six feet four inches of muscle. I *was* surprised about the owner. "I did all my negotiations with a woman, Lisa Arborough," I said. "Was she with him? What did he say his name was?"

"He came alone, and he said he was Brad Arborough. Shall I let the whole thing blow over?"

"I'll handle it from here, Lurleen. I'll let you know."

I filled Mason in on what had happened.

"The man doesn't have a leg to stand on," Mason said. "Lurleen has the rental agreement."

"I know, but I think I should call Lisa and find out what's going on."

I searched my cell for her contact number. She listened as I explained about Lurleen's encounter. I put her on speaker phone, so Mason could hear the conversation.

"I'm so sorry you're in the middle of our domestic dispute—especially when you are about to get married. My husband, Brad, can be a bully. We're negotiating a divorce, and no part of that is going smoothly. I should have known it would be difficult—I was around for his first divorce, and that was a complete fiasco.

"His first divorce?" I asked.

"Long story—everything with Brad is a long story." Lisa paused. "I'm sorry I'm going on and on. It's all so aggravating. We own the houses jointly, and it seems to be a major sticking point as to who will get them in the settlement. Don't worry about whatever Brad said—I'll take care of it. He won't be sending any unexpected guests to mar your festivities."

I sighed with relief. "Thanks. I'm sorry for any extra trouble this might cause you," I said.

"My husband causes me trouble every day—that's why we're getting divorced. But I won't have him ruining your wedding. You have my word on that. Whatever I need to do to keep him out of your hair I'll do. Call me immediately if you or Lurleen hear any more from him."

"We will, thanks."

Then she said the words I didn't want to hear.

"I know it's a small wedding, but will you have security present? You might want to get a local cop to be there just in case. Off duty. They won't charge much. I can give you some names."

"Why? Why do you think we might need that?" I asked.

"I don't think you will. It's simply that my husband has a short fuse. I want to head off any trouble before it happens."

Mason nodded at me.

"I'll take some names," I said.

She rattled them off with phone numbers, and then we hung up.

"This is not what I want for my wedding," I said. "Should we see if we can rent a different place?"

"Everything's booked," Mason said. "We're lucky to have these houses next to each other—there won't be anything else available on Cape Azul. Call a couple of the people on the list, and we'll play it safe."

I called the first name and explained the situation.

"I know Brad," a man named Lenny said. "Can't say I'm surprised he and his wife are divorcing. They used to spend a lot of the summer on the Cape, but in the last few years they never came at the same time. He likes to stir up trouble, especially after a few drinks, but it's nothing I can't handle. Fist fights, that kind of thing.'

"Hire him?" I mouthed to Mason.

"Hi," Mason said. "I'm the groom. I'm also an Atlanta police detective. Can you give me some references?"

He gave Mason the number of the local police department and the name of the local chief of police, Scott Huxley.

"We'll get back to you," Mason said. "You're available tomorrow?"

"Sure, as early as you want."

"Any chance you can come tonight?"

"Off work at five. I can do an all nighter if you want to make sure Brad stays clear of the property. I got a friend who could work as well." He gave us the name of the second man on the list.

"We'll let you know," Mason said and hung up.

He called the station, put his cell on speaker phone, and talked to the chief of police.

"Scott Huxley here. I know Lisa and Brad. Brad can be a pain in the neck. Lisa, his wife, she's fine. I don't know why she put up with him for as long as she did. She's the one who made most of the money—top real estate agent in Atlanta, so she didn't stay around for the money. I think she actually loved the guy."

"You know her well?" Mason asked.

The chief hesitated a beat too long. "Yeah, I know her well, as well as any of the summer residents. She's a straight shooter, and I can vouch for the two guys you mentioned."

We stopped by the side of the road for a picnic I'd packed. Mason made the call to the two police officers. One would come around six tonight and work overnight. He'd be joined by the friend the following morning. Both men would stay through the day.

"We're covered, Ditie," Mason said.

"I don't want the kids or our guests to be in danger," I said, "and I don't want them to worry that we have the place barricaded."

"I told the guys no guns," Mason said. "We'll say they're here to keep the peace, let us know if we're making too much noise, that kind of thing."

"I guess we can live with that," I said.

"No choice, really, Ditie. I don't want anything to spoil our day."

We arrived around four, settled into the smaller house and then joined everyone else next door fifteen minutes later. The kids acted as if they hadn't seen us in a month.

"Mommy," Jason said. "Can we get back in the water? Lurleen said we should wait for you, so we did."

"You read my mind."

The first thing I always did when I arrived on Cape Azul—if it wasn't pouring down rain—was jump in the ocean. That first dip was like anointing my soul. If I could get in the water as soon as I arrived, then whatever happened for the rest of the vacation didn't matter.

Mason and I changed quickly, slathered sunscreen over both our bodies and joined the kids back inside the big house.

Lucie and Hannah were smoothing sunscreen on each of their backs. They stopped long enough to give us a hug. "We have so many surprises for you and Uncle Mason," Lucie whispered. "Anna let us help her make the wedding cake. You won't believe what she put inside."

"What she put inside?" I asked.

"Don't ask me, Mama, because I won't tell you," Lucie said. "You'll see. Tomorrow will be the most perfect day."

I took a long look at both girls. "It looks as if you've already had some perfect days." They looked healthy from the sun and had obviously given themselves manicures and pedicures. Hannah's fingernails were dark blue with moon shaped crescents on each thumb. Her toes were sparkly yellow. Lucie had bright red fingers and toes with what looked like tiny shells added to each fingernail.

"What is all this?" I asked.

"Lurleen helped us," Lucie said. "She told us that if we were going to be bridesmaids we had to look the part."

"Well, you certainly do."

Jason and Noah were waiting for us impatiently near the sliding-glass doors to the veranda. "Can we go now?" Each of them held a boogie board.

"Where'd you get those?" I asked.

"Uncle Danny," Jason said. "He's teaching us how to surf."

I glanced back at Danny who wasn't dressed to join us in the water. "It's Mason's turn," Danny said with a broad smile.

I looked at Mason.

"How hard can it be?" he asked, "especially on a day like today."

He was right about that. The water looked as smooth as a plate of glass and almost as clear.

We headed out to the beach while Danny, Lurleen and Anna stayed in the house. I'm sure they needed a break from all the activity with the kids, and Anna was probably working on dinner. Mason's family was still on the road and due to arrive in an hour or so. Tommy and Josh were in Apalachicola and would be home in time for supper. My boss, Vic, and her husband, Eric, were taking a nap after an earlier arrival.

That meant we had an hour to be in the water with the kids before the evening bustle got underway. I didn't spot a single stingray, but we all wore our water shoes just in case. The ocean temperature was perfect—by my standards anyway—it had to be near eighty degrees.

It was late July, but Cape Azul never got crowded. It really was part of the Forgotten Coast, and with no big attractions, many people still didn't know it existed.

A few shade tents dotted the sand on either side of our houses, and we nodded at the dozen people who lounged inside them as we made our way back from the ocean. By the time Mason and I had changed and come back to the big house, his family had arrived. Mason's mother, Eddy, took care of the introductions.

Brian, whom I'd seen only once before, looked a lot like his dad—same eyes, same smile, solid build. He had more hair, but it was already thinning. Kyle must have taken after his mother—his hair was dark, curly, and he was thin and tall. Neither one seemed to take after their grandmother, Eddie, who was tiny with sparkling blue eyes and a non-stop smile. Both sons hugged their dad warmly and treated me as if I were a welcome addition to the family.

Their wives, Heather and Megan, were equally warm although preoccupied with their tired children. They whisked them away for a nap before dinner.

Vic and her husband came downstairs in time to meet

everyone. I'd never seen her look so relaxed. "This is wonderful," she whispered to me when we were briefly alone. "You know what a workaholic Eric is. Only your wedding could drag him away from his research. Eric was a scientist in biomedical research at the local university in Atlanta.

"That might be the pot calling the kettle black," I said to Vic.

She laughed. "This break is good for both of us."

Anna prepared a buffet meant to be eaten when people arrived and were hungry. Mason and I settled on the veranda with the kids and helped ourselves to pasta, salad, cold shrimp, and a glass of white wine. People came and went as they wished. Every one of them told us how happy they were about our upcoming marriage.

We were seated on the veranda with ice cream sundaes when the sun began to set. Spectacular sunsets were a feature of Cape Azul. The peninsula faced west, and we all grew quiet as the sun began its descent. I'd always wanted to see the green light some people claimed to see when the sun dipped below the horizon. I never had before. This time I saw it—a single flash of green as the sun disappeared. Gone in a second.

"A good omen," I said to Mason.

"A very good omen," he agreed.

3

I woke early the next morning and left Mason to get his beauty rest. He was a lounging-in-bed kind of guy when he wasn't working. I put on my walking shoes, a little sunscreen for when the sun came up and headed out the door.

A police officer stood at the foot of the stairs. He looked good in his uniform but far too young to be a real police officer, more like a teenager dressed for a Halloween party with longish blond hair appearing beneath his cap.

"You're Lenny, right?" I asked.

He nodded and straightened his jacket. "The Chief thought I should wear this, so anyone who saw me wouldn't think I was casing the joint.

"Any problems last night?" I asked.

"None. I'll wait for my partner to arrive and then I'll take a nap. We'll both be on duty before the wedding."

"Thanks."

The sky was clear and the waves broke gently on the shore line. It looked to be high tide. No one else was around for the first part of my walk. As always I scanned the horizon for dolphins. Nothing this morning.

I let my mind drift. We could have the wedding on the

beach as planned. I hoped our marriage would be as care-free, but I knew we'd run into problems. Every marriage did and we'd have to sort them out. Mason and I understood our quirks, our stubborn points and our commitment to always have each other's back, to always love and care for Jason and Lucie.

The world brightened suddenly, the way it usually did when the sun popped up from behind the dunes. I inhaled the fresh salty air, perfect for the start of our new life. Sandpipers scurried along the water's edge. A few plovers hopped about higher up on the beach and laughing gulls cawed overhead. I needed Lucie with her notebook to teach me about all the birds on the gulf not to mention the shells. I knew this would happen over the years to come.

So many wonderful adventures lay ahead of us.

I glanced back toward our rental house and saw a figure approaching. Not Mason. This figure wasn't male, and it wasn't anyone I recognized.

I waited for the person to catch up with me. She was a woman in her late forties I'd guess with platinum hair, a trim figure, and a broad smile full of straight white teeth.

"Hi," the woman said. "I'm Lisa, Lisa Arborough. We spoke yesterday—you and your friends are staying in my houses. Lenny said I might find you here."

"The houses are beautiful," I said. "What brings you out so early in the morning?"

"I needed to walk off some nervous energy, and I wanted to make sure you had everything you needed for your big day. I hope you stay happily married—divorce proceedings are not for the faint of heart. It feels as if we're relitigating the first divorce, the one that really destroyed the Arborough fortunes. Brad's ex-wife, Shannon, and his younger brother, Howie, are in town and seem to be determined to get some money out of our divorce. I'm not sure how they expect to do that. There isn't much money to get, but Shannon is still furious with me

—just because Brad and I were involved when that divorce finally happened."

Lisa stopped talking and turned towards me.

"I am so sorry," she said. "All this on your wedding day. When I get angry, I talk too much. Really, I only wanted to make sure Brad hadn't bothered you again. Lenny said he hadn't seen Brad, but I wanted to check with you in case he'd called."

"No," I said. "He hasn't called."

"I'm delighted to hear that. I couldn't reach him yesterday, and that worried me a bit."

She seemed reluctant to leave. "Do you mind if I walk with you?"

"Not at all."

"I love walking on the beach in the early morning," she said. "I also love the house you're staying in. I may decide to live there permanently once Brad and I divorce. Anyway, my attorney spoke to his attorney and convinced Brad to stay away if he wanted me to negotiate in good faith. I guess he's living up to that agreement."

"You must be happy about that," I said.

"Hmm, I should be."

She didn't sound happy. I looked at her.

"I don't know what will happen next," Lisa said. "Brad and I have had a complicated relationship for the ten years of our marriage. I ought to be happy it's finally ending, but we really did love each other in the beginning. He claims he never loved me, but that's what men always say when they're ready to move on."

"I thought you were the one moving on," I said. "The chief of police said you'd finally had enough."

"Scott Huxley? Scott is a sweetheart. His wife divorced him a couple of years ago, and I think he hopes . . . well, he hopes I've finally had enough and might soon be available. But you should know gossip around here spreads like wild fires in California. People get bored and have to invent a story. Brad was the one who wanted out."

She stopped and stared out at the ocean. "I'm almost fifty years old and I don't want to start over."

"Are you hoping for a reconciliation?" I asked.

"I'm afraid it might be too late for that."

I couldn't think of anything encouraging to say.

"I've taken up too much of your time," Lisa said, "and on the day of your wedding. I'm sorry."

"I'm sorry this is a hard time for you," I said.

"I'll get over it. I'll leave you here." She turned and headed back the way she'd come. I looked at my watch. Time for me to head back as well.

Lisa Arborough strode across the sand. She spoke as if she were a broken woman, but she walked as if she were very much in charge of the situation. The story made me curious, but not for long—not on my wedding day.

I came home to find Mason on the lower deck, watching out for me. Beside him was a table with what looked like mimosas and some of Anna's pastries.

He kissed me. "You look radiant. Anna brought these over for us. Rugelach. She said it's never too early for something sweet and a little champagne on a person's wedding day."

I told him about my conversation with Lisa.

"She seems sad about the divorce," I said. "It's confusing. I can't tell if she wants to go through with it or wishes she could stay married to him. She said she hadn't been able to reach Brad yesterday and was glad he hasn't caused any more trouble."

Mason and I sat side by side sipping mimosas, eating Anna's spectacular rugelach and discussing details for the upcoming ceremony. Lurleen promised that she and Eddy would take care of everything including making sure the kids were dressed appropriately. All we had to do was take care of ourselves.

"Ready?" I asked, glancing at my watch. "Nine-thirty."

Mason and I went upstairs to separate bedrooms to change. Lurleen popped over at nine forty-five. She looked beautiful in a simple sleeveless summer dress of periwinkle blue. It hugged her slim figure and complemented her copper curls.

"Eddy has everything under control at the big house, and I couldn't let you get dressed without my help," she said.

"Thank you," I said.

Lurleen zipped up the back of my A-line wedding dress. It was covered in lace that made me feel beautiful with a white shimmering chiffon cape that draped over the dress.

"Now for your hair," Lurleen said.

"You can't do anything with my hair," I said. "You have to let it be. I keep it short because that's the only way it will remotely behave."

"I'd never do anything to your hair," Lurleen said, "except to add a band of flowers—snapdragons and freesia."

"My favorites," I said.

"I know! The florist attached them to this silver band," Lurleen said. "Look at yourself in the mirror."

I did. "I feel gorgeous."

"You are gorgeous," Lurleen said. "Here's your bouquet—wildflowers. Lucie and Hannah will be carrying similar but smaller bouquets."

"Perfect."

"I'll carry the ring for you to give to Mason," Lurleen said. "Danny has yours."

Mason and I had gone together to have two simple rings made. Mine would look beautiful with the emerald engagement ring he gave me.

"I think you're almost ready," Lurleen said. "Now, here is something borrowed."

She handed me a charm bracelet that I'd never seen

before. On it were five charms—a book, an ice cream cone, a little girl, a mother, and a bicycle.

"My mother gave this to me when I was seven and added a charm for each of my birthdays until she died. It's the one way she could show me she loved me despite the drugs and crummy boyfriends."

"Oh, Lurleen," I said. "Thank you."

"Now for something blue," she said.

I showed her the pinkie on my right hand. "The girls already handled that. Hannah painted my pinkie to look exactly like hers, complete with a crescent moon."

We both smiled at that.

"You've got something new—your dress. What about something old? You could count the charm bracelet, I suppose."

I pulled a pendant from around my neck and showed her a small medallion that hung on the end. "This was my dad's. He worked as a policeman in Iowa before he met my mother. He got this as a commendation when he rescued a child and her mother from a car that flew off the road into a lake. The car was underwater when he reached them, but he managed to save them both."

"Perfect," Lurleen said. "You never told me that story." She pushed back her wavy hair to take a better look. "So he was a brave man as well as being kind and loving."

"He was all those things," I said. "A lot like Mason."

I slipped the pendant back under the neckline of my dress, and walked out on the deck.

"People are already on the beach. What's that?" I asked, pointing to what looked like flowers in the sand.

"That is your circle of love," Lurleen said, "where you, Mason, and the minister will stand as you say your vows."

"You've thought of everything, Lurleen," I said and hugged her.

She blushed. "That's what best friends do. Now, are you ready?"

"Completely ready."

I slipped on my satin sandals and we headed outside. Our minister came up to me.

"Any last minute instructions?"

"No," I said.

I walked over to Lucie and Hannah. "You two look beautiful."

"Not as beautiful as you, Mama," Lucie said.

Hannah nodded.

The girls had already done their part, strewing rose petals along the sand to Lurleen's circle of love. Mason stood inside the circle, grinning from ear to ear.

"You look so beautiful," he whispered.

"So do you," I said. Mason wore a summer suit of white linen with a blue cravat that looked like the ocean behind him.

The only thing we lacked was music, but I liked the quiet space. Later, there would be plenty of music. All we heard now were the waves lapping on the shore and the distant laughter of children in the water.

As the minister started speaking, our guests gathered around us in a semi-circle as if to say they would always protect us. Inside the circle stood Jason and Lucie, hands clasped, beside Mason and me.

Like most brides, I have no idea what was said or how long the ceremony lasted. I remember exchanging our rings, and I remember the minister saying we could now kiss one another as husband and wife.

Bliss.

We walked back to the big house where Anna had brunch set out for us. All my favorites and Mason's as well. Eggs Benedict, biscuits and southern gravy, a fresh fruit salad, bacon, sausage and omelets for anyone who wanted one. We ate and groaned that we would never eat another meal.

Then it was time for cake—a lovely, simple sheet cake on which sat paper mache images of me, Mason, Lucie, and Jason. Hermione and Majestic sat on either end of the family.

"How in the world?" I asked Anna.

"It wasn't me—that is purely Lurleen. She also had a say in what went in the cake."

Lurleen beamed. "That's why we needed a sheet cake—I hope you don't mind. I had to have room for my creations and I had to know who was getting each piece of cake."

We'd done this once before—hidden treasures in a cake to foretell our fortunes. It seemed Lurleen knew precisely how to slice the pieces, so everyone would get their appropriate fortune.

"Be careful," she advised us. "Look for the wax paper and unwrap it—don't eat it."

People oohed and ahed over their treasures that represented all kinds of good things coming to them. Lucie got a tiny piano. Hannah, a little house full of people. Jason got a curious miniature puzzle, which Lurleen explained was for that problem-solving mind of his. Noah got what looked like a gold-plated soccer ball. Lurleen gave Mason and me one special slice of cake. "Your fortunes are one. Here you go."

Inside, we found two silver hearts intertwined. "That's the two of you forever," Lurleen said.

And I knew it would be.

By early afternoon. we'd finished our wedding festivities.

"People are free to do what they want with the rest of their day," Mason said. "Ditie and I will retire to our private retreat. Let's meet again in the big house around six."

Mason and I changed out of our wedding clothes, fell into bed and slept for an hour. Then we woke and shared our favorite moments from the morning. We spent the rest of the afternoon in bed but not sleeping.

Dinner was an outdoor affair on the veranda. Music and

dancing were inside, doors closed, curtains drawn so we wouldn't disturb our neighbors—or the sea turtles.

By ten p.m. we were all exhausted. Even Danny and Lurleen seemed to be ready to call it a night. Vic and her husband thanked us for a wonderful time. They planned to head home early the next morning. Josh and Tommy also said they'd be leaving early the next day—Tommy had a big court case coming up and Josh had his patients to worry about. We said goodbye to them and thanked them for coming.

"You've given us lots of good ideas about how to do a proper wedding," Josh said, "when Tommy finally decides he's up for it."

We'd already said goodnight to the kids. They were in their respective rooms eager for a sleepover with their best friends. Of all of us, the kids seemed to be the least tired. We told them how much we'd enjoyed sharing our wedding with them.

We thanked Anna for her fabulous meals, then Danny and Lurleen for everything else. Mason and I walked back to our own house. It was lights out everywhere to protect the sea turtles that might be crawling in to lay their eggs. That was the Florida ordinance. The sea turtles could get confused by lights from houses and lay their eggs in all the wrong places.

"It's been a wonderful day," Mason said. "Are you as happy as I am?"

"Happier," I said.

"Not possible."

4

Mason and I slept in. Anna brought us breakfast in bed after she called to make sure we were awake —sometime after ten.

"What's going on next door?" I asked.

"I saw the kids go out with Danny pretty early this morning," Anna said. "They took a walk and then they went for a swim. Tommy and Josh left around nine. They said to tell you it was the best wedding they'd ever been to with the best food. I sent them on their way with rugelach and they claimed it was the best pastry they'd ever eaten. Those boys know how to flatter a girl."

"That *is* the best pastry, Anna, next to your cinnamon rolls," I said. "That wasn't flattery."

Anna blushed. "All this praise is a little overwhelming for me. Would you like me to stay another day and fix you dinner?"

"We have it covered," I said. "You've been wonderful. We are so happy you agreed to come to Atlanta and manage Lurleen's cooking school. Are you missing South Carolina?"

"I miss Beaufort a little, but Atlanta is beginning to feel like home now. Lurleen lets me do what I want with the school. I can try out new recipes, and I've had the most wonderful students— adults and children—including Lucie and Hannah. I couldn't be happier. We may open the dining room for special catered events, and we're continuing to train chefs through our scholarship program."

"That's a part of the school that Lurleen really cares about," I said, "helping people get a leg up in life. Now, you are off the clock. Leave whenever you want—just take your own stuff—your knives, special pans. Lurleen and I will bring the rest later."

Mason and I drank our coffee, ate our rolls and then dressed slowly to welcome the day. The kids found us on our deck downstairs an hour later.

"Mommy," Jason called out to us. "I thought you would sleep all day. Lurleen says we have to move down the beach this afternoon, and I thought we wouldn't get to see you before we went."

"We'd never let that happen," I said.

I hugged him and Mason did the same.

"Are you and Noah having fun?" Mason asked.

'Yes. Danny runs with us on the beach. He's fast but sometimes we're faster."

I had to smile at that. Noah and Jason were eight-year old soccer players, full of energy. Danny was over thirty—big and muscular. I wasn't sure if he let them win or if he couldn't quite keep up with them.

"Where is Noah?" I asked, looking around.

"He's in our room watching TV. Danny said he's on vacation and we should let him do what he wants. Why do we have to move down the beach?"

"I guess so we can do what *we* want," Mason said. "Grown-up stuff. We're on vacation too."

"We'll see you every day. I promise," I said. "You'll only be half a mile away."

"That's what Danny said. He said we'd have a lot more

fun with him and Lurleen than with the two of you. He said people who just got married got kind of lazy and didn't run around a lot."

Lazy, I thought. That's one description. I grinned at Mason.

Lucie and Hannah joined Jason on the deck.

"We're about to leave, Mama," Lucie said.

Since the adoption, Lucie no longer called me Mama Di—now I was strictly Mama. It didn't mean she'd forgotten Ellie. That would never happen. Ellie was her first mom, and she'd always call her Mom. Her death was something we'd be dealing with forever. But now, Lucie was happy to have a second mother and soon a new father. I had no idea what the kids would call Mason.

Hannah reached out her hand to me as if she meant to shake it. I hugged her instead.

"You were so beautiful yesterday," she said. "Thank you for letting me come."

"You're family now, Hannah," I said. "You had to be here."

Over the last year, Hannah had transformed from a lonely, petulant twelve year old into a warm and loving teenager. She spent as much time at our house as she did at her own. I knew there would likely be turbulent times ahead for both the girls—teenagers were rarely easy. I knew that as a pediatrician. I also knew it from my own rebellious teenage years. With Mason beside me, we'd make it through.

Lurleen came last. "We're all packed up and ready to go. Anna insisted on leaving us dinner for tonight. This is your honeymoon and the kids understand that. You can have as much private time as you want."

"Thanks, Lurleen. We will want to see them every day—but maybe tonight, Mason and I will eat in. We'll keep you posted."

"What about your boys and their families?" I asked Mason when we were alone once more.

"They plan to stay one more day and then they'll spend time together in Atlanta at Eddy's. They haven't seen much of each other recently, and they want their kids to become close cousins as they grow up."

"We have the big house for another day, so that will work," I said. "Much as I loved all the partying, I'm really looking forward to time alone with you."

"Me too." Mason pulled me into a long embrace. "Time to go back to bed, I'd say."

"No reason not to."

We loved the quiet and the closeness for the rest of the day—until the banging on our front door started at 3:00 p.m. It didn't stop until Mason pulled on his robe and ran downstairs.

"Who is it?" he yelled through the closed door. I stood behind him.

"Your renter from next door—Wayne Baskin. I was supposed to move in at three this afternoon. It's three now and the place is still occupied. Brad said he'd told you I was coming."

So much for Lisa Arborough taking care of things.

5

Mason didn't open the door. "Give me one minute," he said. "I need to change and then we'll sort this out, Mr. . . . "

"Baskin. Wayne Baskin.Brad didn't tell you I was coming? He told me to knock on this door if there were any problems about getting in."

"We're renters too, Mr. Baskin. We made our arrangements with Mrs. Arborough."

He yelled all this through the closed door.

"Did Lisa Arborough contact you?" I asked. "She said there was some mistake and she'd take care of it."

"No one called me except Brad. He told me the place was mine starting today for as long as I wanted it."

I grabbed Lisa Arborough's card off the coffee table and opened the door wide enough to hand him the card.

"Call Mrs. Arborough. She'll settle this with you. The big house won't be free until tomorrow."

Wayne Baskin scowled at me and the card. "I know her number," he said, handing the card back to me.

If not for the scowl he would have looked like a pleasant-enough man. He had laugh lines around his eyes, although he wasn't smiling at the moment. Even when he'd raised his voice, he hadn't seemed particularly frightening. He left without another word.

Mason and I dressed quickly. "I better check next door," he said, "and see if he's been bothering my sons or their wives. You stay here."

"Right."

I drank a cup of coffee and waited near the front door. Wayne Baskin got back to me before Mason did. He pounded on the door the way he had before as if he meant to raise the dead.

I opened the door but left the screen door latched.

"I can't reach either one of them," he said. "I've got my stuff in the car, and I don't want to waste all this time. You got any other ideas?"

I shook my head. "I don't know Lisa. I only met her yesterday. She said her husband had made a mistake when he told you that the place was yours. She promised she'd work it out."

"She never worked it out with me. Never called. Nothing. Now, what am I supposed to do? I drove down here from north Georgia."

"I'm sorry, Mr. Baskin. My husband will be here in a minute. We'll do what we can to help you sort this out."

That seemed to settle him down. I invited him inside and offered him a cup of coffee and one of Anna's pastries. It would be difficult for anyone to remain angry while eating Anna's rugelach.

"These are as good as the doughnuts they sell down the road," he said. "Look, I'm sorry I came on so strong. I know it's not your fault. Probably not Lisa's either. Brad thinks he can get his way about everything."

"Brad is a friend?" I asked.

"Used to be. I hadn't heard from him in months. Then he upped and called me out of the blue. He owes me a lot of money and said he was ready to make good on what he owed me. He promised we could make some kind of deal that would let me get my money back. Said I could stay in the house for free, maybe give me the smaller house in compensation for the debt. That's why I came down here. Brad was always a fast talker. I should have known better than to trust him."

"Did you know he and his wife were getting divorced?" I asked.

He shrugged and shook his head no. "Can't say I'm surprised though. I'm more surprised the marriage lasted as long as it did. Lisa could have had any man she wanted. Why she chose Brad I'll never know."

"You aren't the first person to say that."

Wayne Baskin just nodded. "Got any more of these?" He pointed to his empty plate. "I haven't had anything to eat all day."

I brought him out a plate of rugelach and poured him another cup of coffee.

Mason returned as Wayne Baskin finished his second roll.

"Mr. Baskin came back when he couldn't reach either of the Arboroughs," I said.

"Not sure what to say about that, but the boys and their families will be leaving today." He turned to me. "They claimed that was their plan all along. I'm not sure what to do about you, Mr. Baskin. We don't own the place, so we can't let you stay there."

"Let me try Lisa once more," I said. I left her an urgent message—that Mr. Baskin was here expecting to stay in the big house and that our guests were about to leave. What did she want us to do?

The reception was always spotty on the Cape, but I managed to get enough bars to text her from the deck.

She called back twenty minutes later.

"I am so sorry," she said. "I haven't been able to track

down Brad. I know Wayne. He can stay in the big house once your friends leave. I'll see if I can get a cleaning crew in there this afternoon or tomorrow. Is he with you? Maybe I should talk to him."

Lisa sounded distracted. She seemed to be calling from some place outside—I could hear wind in the background.

"Are you all right?" I asked.

"Yes, of course. Put Wayne on the phone, please."

I did that. We could only hear Wayne's side of the conversation.

"No. I tried to call him once I arrived and found people still in the house . . . no, I haven't heard from him in a week. The boat's gone? You know Brad—he's probably living on it or sleeping off a party he had . . . . Can't say I'm sorry about the divorce, Lisa. You deserve a lot better . . . . I'll call right away if I hear from him, and thanks about the house."

Wayne Baskin handed my cell back to me. "She cleared it. I can move in next door as soon as your folks move out. Thanks for your help—I'll wait in my car."

"You don't have to do that," I said. "It's too hot. You can wait here—inside or on the deck."

"I'll sit on the deck so I don't disturb you. You've already been too kind."

I was beginning to warm up to Wayne Baskin. Mason said he'd go help his family pack and let me know when they were ready to say goodbye. I led Wayne out to the deck. It was after five, so I offered him a glass of wine.

"Thanks, but I don't drink anymore. That's how Brad and I met—in a bar. We were great drinking buddies, but it got to be too much for both of us. Destroyed my marriage. It probably destroyed his as well although I knew he and Lisa had all kinds of other problems. Brad blamed everything on her. He said he couldn't get a job because of her—figure that one out. He claimed she destroyed his self-confidence—she was such a go-getter. Of course he used a different term than that. He said she was the reason he drank—always criticizing him, calling him lazy and a liar. I asked why he didn't leave her if that was how he felt. You know what he told me?"

I shook my head no.

"He said he couldn't survive without her money. He told me he couldn't go back to being poor. That's what he said. First honest thing he probably ever said to me. Why she wasn't the one wanting out—that I'll never know. There were plenty of men interested in her."

"You sound as if you might have been one of them," I said.

He looked at me, surprised. "I guess I was, but I knew Lisa could never be interested in me. Look at me."

I studied him. He struck me as a good-looking man in his early fifties. Short brown hair—not quite a buzz cut. Glasses. Twinkle in his eye. Trim as if he played tennis or skied.

"I'm not sure what you think I see."

"Thanks. I don't know if you're being kind. I'm not a bad sort, but Lisa and Brad liked to party. When I stopped drinking I stopped partying as well. I'm a quiet guy."

I had to smile at that. "You didn't sound quiet when you were pounding on our door."

"Oh, yeah. Sorry about that. I was frustrated. Brad said I could use the house as soon as I wanted to. He owes me thousands of dollars and said he'd be around to get that debt settled in the next week or so. I should have known better. Brad was always a big talker who didn't follow through on much of anything. He had these schemes that he said would make us both wealthy."

"What made you give him money in the first place?"

"I'm a soft touch. Or I used to be when I was drinking. Brad said he needed the money to get started in a PR business promoting Apalachicola as the new Dollywood. He convinced me he had contacts, knew all about the entertainment business, and that Apalachicola was going to be the next music hub, bigger than Nashville. Brad could sell anything to anyone. He certainly sold me. He promised me a piece of the action or my money back with interest. I finally opted for the money when nothing changed in Apalachicola. That was a few years ago. I haven't seen a cent so I came down here to settle a score."

My cell rang. I picked up without looking at the number. I assumed it was Mason or Lurleen.

It was Lisa. This time she sounded desperate. "Is Wayne still there? I have to talk to him."

I handed Wayne the phone and stood close enough to hear both sides of the conversation. "They found Brad's boat adrift at sea. No one was on it. I'm scared, Wayne. I think something must have happened to Brad."

"Where are you now?" Wayne asked her.

"At the police station. Can you come?"

"I'm on my way."

He handed me back the phone and recapped the conversation I'd just heard.

"Something may have happened to Brad. I have to go. I'll stop back later for a key to the house next door unless Lisa has one. I'll let you know."

6

Mason found me moments later to say everyone was ready to leave. I filled him in on Wayne's sudden departure.

"That isn't a great sign," Mason said, "a boat adrift with no one on board."

"No, it's not," I said. "Lisa sounded worried—she hadn't spoken to him in a couple of days—and Wayne described Brad as a big drinker. I hope he didn't fall overboard."

Mason took my hand and gave me that look of his, the one he sometimes gave to Lurleen when she was telling a fantastic story that might end in disaster.

"I know," I said, "don't borrow trouble."

"It's not that so much," Mason said. "People drink on boats and all kinds of tragedies happen because of that. It's more that there isn't anything we can do about it. We can only sit tight and hope for the best."

We walked over to the house next door to say goodbye. I liked both sons and their wives and wished we had more opportunity to visit with them. Brian, the younger one, gave me a bear hug as soon as I walked in the door.

"Beautiful wedding. Dad is so lucky to have you."

His wife, Heather, nodded along. "You've brought out the best in him. We are all so happy for both of you." Heather was a beautiful girl, dark hair, tawny skin, a little taller than Brian. "Have you heard we may be moving to Atlanta? If I play my cards right, I think I can get a transfer there—I work for Amazon. Brian says it's my turn to make the career move, so he'll follow me. As a high school teacher and coach, he won't have trouble getting a job in Atlanta."

"That sounds wonderful," I said. "Then I can really get to know you both, along with that beautiful daughter of yours —Mia."

"Lucie did a great job with the kids. What a wonderful girl!" Heather said.

"Yes," I agreed.

Kyle and Megan came up as Heather was talking. "I wish we could move closer, too," Megan said, "so Olivia could grow up near all of you. Kyle has a new assignment in Indonesia. I'm not complaining—I love that part of the world. It's just so far from family."

"It's a two-year assignment," Kyle said, "and after that we might be closer to home. Count on us for Christmas."

"Wonderful," I said. While Kyle didn't look much like his dad, Megan and I could have passed for sisters. She had the same full figure I had and curly dark hair. Her hair seemed a little more responsive to her efforts than mine did, and she wore it long, pulled back in a low ponytail.

We said good bye to the grandkids as they got strapped into their car seats. Then we offered to take one last look through the house.

"Don't bother," Megan said. "If you find something valuable, give us a call and we'll figure out how to get it from you. Otherwise you're welcome to anything we've left behind. We travel light—or as light as you can with a baby—so no worries."

Eddy was the last to come downstairs with a small

overnight case. I tried to take it from her but she shook her head. "Weight training," she said and laughed. "It's been delightful, but I have to admit I'm ready to head home. I want to get back to Snickerdoodle. That dog loves his kennel—I believe he considers it his spa—but I miss him."

"You'll have a full house," I said. "The kids will stay with you?"

"Absolutely," she said, "and I'll love every minute of it."

She took my hand.

"I can't tell you how happy I am that this moment has arrived. I knew from the first time I met you that we were intended to be family. And now we are."

"I feel exactly the same way," I said.

She hugged me—a long, warm hug that said I was now part of her family and she had no intention of ever letting me go.

"We'll stop to say goodbye to the kids before we leave Cape Azul," she said.

One last kiss on the cheek and she headed out the door.

Mason and I stood on the back porch and waved goodbye as they drove off. I felt like crying. I never liked seeing people leave even though I knew I'd see most of them again soon.

"Handkerchief?" Mason asked.

"No," I said. "I'm fine."

"We probably should go through the house now that everyone's gone and make sure nothing's been left behind," Mason said. "You up for that?"

"Sure."

"We'll take a walk through, and then I'll concentrate on the upstairs," he said, "and you can take the bottom floor."

Our guests, family really, had left things in good shape. They'd stripped the beds, put a load of towels in the washer, and thrown out the trash. I wasn't sure when the cleaning crew was coming, but they wouldn't have a whole lot to do. One room, the true master suite, looked as if no one had stayed in it. The sheets were still on the bed with the towels folded neatly on top."

"It looks as if no one used this room," I said.

"My kids never wanted to upstage one another," Mason said. "I'll bet no one was willing to claim what was obviously the best bedroom in the house. It's a great house with so many rooms."

I smiled and nodded. Then I thought about how much friction that house had already caused for Lisa and Brad . . . and us.

I felt the smile freeze on my lips. I remembered the issue of Brad's empty boat at sea. "I wonder if I should call Lisa," I said.

"Let her call you," Mason said. "If you haven't heard from her by tomorrow, then we'll call. Or most likely, Wayne will know something more when he returns. Let's finish going over the house and maybe the mystery of the missing husband will be solved by then.

I was running the dishwasher when my cell rang. It was Lisa.

"Hello, Dr. Brown. I'm sorry to disturb you. Wayne said you'd been really helpful to him, and I'm hoping you can do me one more favor—let the cleaning crew in when they arrive. They promised to be there in half an hour."

"Of course."

Lisa sounded calm, like it was business as usual.

"We heard there was a problem with your husband's boat—is it resolved?"

"I warned you news travels fast on the Cape," she said. "It's not resolved. We don't know where Brad is at the moment, but he's most likely in some bar. Apparently, some kids managed to take that boat out into the water last night. They claimed it was unmoored when they found it and they tried to jump ship when they saw a Coast Guard boat coming to check on them."

She paused.

"I'm not sure what made me so worried about Brad, except that the boat is the one thing he seemed to really care about. I promised he could keep it in the settlement. He always said he intended to die on that boat. Anyway, I hope he'll show up for our second mediation session. The first one didn't get us very far. I'm sure that's why he called Wayne down here—he was trying to establish he had some claim on the house, and he put Wayne in the middle of it."

"We're here in the house right now, so we'll wait on the cleaning crew, no problem."

"Again, I am so sorry to bother you on your honeymoon. I'll plan to come over later to pay them, check their work and get Wayne settled. I should know by then where my missing husband has landed."

We hung up. I finished checking the downstairs for any forgotten items. I found one puppet—a baby dinosaur that popped out of an egg. Someone might be missing that. I called up to Mason. He came downstairs with a strange expression on his face.

"Good thing we looked," he said. "See what I found?"

He showed me an emerald and diamond bracelet in a box.

"Is it real?" I asked.

"Looks real to me," Mason said. "Neither of my boys could afford something like this. I mean I suppose it could be an anniversary present but both their anniversaries are months away. It was in a box tied with a ribbon as if it were meant to be a gift."

"Where did you find it?" I asked.

"In the unused master bedroom, tucked under a pillow. I'm not sure what made me enter that room, but I saw the ribbon sticking out from under a sham—right word?"

"Right word," I said. The emerald and diamond bracelet was gorgeous. For a moment I wondered if this was a surprise gift from Mason. Before I could say a word, he shook his head.

"A bracelet like this would have cost me a year's salary, and I know you would have wanted me to spend that money on something more practical."

That was the truth. I put the bracelet back in the box and handed the box to him.

Mason called his sons who knew nothing about an expensive bracelet. I couldn't imagine Josh or Tommy having much to do with an emerald and diamond bracelet, but I called them anyway.

"I love you, sis," Tommy said, "but that doesn't extend to giving you a bracelet like the one you describe. Besides, you don't wear much jewelry. That didn't come from me or Josh."

I checked with Vic and Anna, but neither of them could account for the bracelet.

The only person left to call was Lisa. Perhaps former renters had left it behind, but surely they would have come back for it or they would have asked Lisa to find it for them.

She picked up on the first ring. She spoke before I could.

"They found Brad! He's dead! His body was washed up on the beach—less than half a mile north of where you are. I'm coming out with the police. Brad was staying in that house until I rented it to you for the wedding."

"I am so sorry," I said.

"Thank you," Lisa said. "Wait, you called me. What is it you wanted to tell me?"

"We found something odd in the house that didn't belong to any of us," I said, "an emerald and diamond bracelet."

"Oh, God," she said. "He's been at it again."

7

"At it again?" I asked.

"Such a long story. If he wasn't dead I might kill him myself," Lisa said. "Sorry, I don't mean that—it's all such a shock! I can't believe he's dead, and I can't believe he used my money to buy a gift for one of his paramours."

Paramours? I mouthed to Mason.

"I'm sorry," Lisa said again. "Ignore me—I'm distraught. I don't know what I'm saying. We'll be at the house in five minutes. Where did you find it?"

Mason took the phone from me. "I found it under a pillow in the master suite—part of a ribbon was poking out. I'll stick around and show the police exactly where I found it."

"Thank you," Lisa said and hung up.

"You can go home, Ditie," Mason said. "I'll get there as soon as I can."

"I'm not going anywhere."

"I'm glad the kids aren't here," Mason said. "They won't have to hear about the body. If Brad is someone who drinks as much as people say, maybe he fell off the boat and drowned.

Boating accidents happen all the time from too much alcohol."

"I hope you're right. If this is more than an accident," I said, "Lurleen will want to know every detail. And that means Lucie will hear about it as well. Sometimes, Lucie and Lurleen are like two peas in a pod—they can't leave any mystery alone."

I took a picture of the bracelet. I knew Lurleen would want to see it. She might even know how much it was worth, although I wasn't sure what she could tell from a photograph.

We sat in the living room of the big house waiting for the police to arrive. Mason didn't have to tell me not to touch anything. I knew the routine, but this wasn't a crime scene, and the house had been lived in by more than a dozen people. It wasn't at all clear what the police might find beyond a very expensive diamond and emerald bracelet.

Lisa arrived with a man in plain clothes. "This is Scott Huxley, chief of police for Cape Azul. Dr. Brown and Detective Garrett."

We shook hands all around.

"We talked on the phone, so I feel as if we've already met," Chief Huxley said. "My men work out for you?"

"Absolutely," Mason said. "Thanks. The wedding was great."

"I'll let you get back to whatever you were doing as soon as possible," Chief Huxley said. "Tell me about the bracelet."

Mason handed the box to him. "There was a ribbon on the box—I left it upstairs. I can show you exactly where I found it."

Lisa peered over Scott Huxley's shoulder as he examined the bracelet. She asked to look at it when he was finished.

"This is beautiful," she said. "More than what he usually gave his girls when he was done with them. I remember what he gave his first wife to soften the divorce—it was also a diamond and emerald bracelet, but it was nothing like this. I remember Shannon threw it back in his face. She wanted a lot more from that divorce than what she called a crummy peace offering."

"You told me she was back in town," I said," hoping to get something out of his second divorce."

"There's nothing more to get honestly," Lisa said. "I think she just wants to cause trouble—she's always hated me."

Mason gave me a look that said none of this was our business.

We followed the Chief up the stairs into the lavish master bedroom. Mason lifted up the pillow. "The end of the red ribbon was hanging out—that's what I spotted."

Chief Huxley put on gloves and examined each sham on the bed. He found nothing else. Then he pulled back the coverlet and the top sheet. We all saw the same thing at the same time—a rust colored stain high up on the bottom sheet, close to where a person's chest might be if he or she were lying on the bed.

"Blood?" I asked.

"Could be," Chief Huxley responded, "but there isn't much of it. No one bled out on this sheet."

"Can you tell anything more, Scott?" Lisa asked.

"Not without forensic folks to take a look at this. And we don't have forensic folks here. I'll have to get them from Tallahassee to do it right. This room is off limits until further notice."

He shooed us out the door. Lisa, Mason and I walked downstairs.

"It's a small amount of blood," Mason said. "Could mean something or nothing at all."

Lisa was quiet as Mason spoke. "It means something," she said when Mason finished talking. "It means Brad was in that bed with a woman. He liked his women passionate, so there was always a little scratching, a little blood. That was his MO. Then, when he got tired of them, he gave them a trinket—a goodbye gift to make sure there were no hard feelings."

"I'd hardly call that bracelet a trinket," I said.

Lisa laughed—the kind of laugh that reminded me of icicles breaking. "Brad was very generous with other people's money, mine in particular."

Scott Huxley joined us downstairs.

"Find anything else of interest?" Lisa asked.

"I shouldn't be telling you this," Chief Huxley said, "but you'll hear it one way or another."

"Right. So what else did you find?" Lisa asked.

"Long blond hairs," he said.

"Not mine, of course." She patted her own short hair. "Any of your guests have long blond hair?"

I shook my head. "No blonds among the adult women, and the men all have short hair."

"Then the guest was one of Brad's," Lisa said. "If there was foul play, I'd be looking for an attractive blond, no older than thirty. I aged out on Brad years ago. Maybe she got ticked at him when he decided to move on. Maybe that's why she never got her goodbye gift from him."

"That's a lot of speculation, Lisa," the Chief said. "We won't know what happened to Brad until we hear back from the coroner, and that could take a couple of days. The coroner is in Tallahassee like the forensic crew. We'll have to be patient."

"And what about Wayne?" Lisa asked. "He was planning to stay in the house."

"Not possible for now," Chief Huxley said.

"Then he'll stay with me. He left to buy groceries. I'll call him."

Chief Huxley had an expression on his face that didn't look happy about that decision.

"I have no choice, Scott. This way you'll know where to

find both of us. Maybe Brad did fall off his boat, drunk, but I won't be surprised if he was murdered. He hurt a lot of people with one scheme or another. Wayne and I won't be the only suspects. You didn't like Brad anymore than the rest of us, Scott, if we're all going to be honest about things."

8

"Need anything more from us?" Mason asked.

"No, you're free to go," the Chief said. "You're staying around for a while?"

"All week."

"Good. You never know when a big city police detective might come in handy," the Chief said.

Was he being sarcastic? I looked at Mason.

"You know where to find me. My wife might be the one you want to talk to. She and her friend Lurleen have solved a few murders in the past."

"Is that a joke?" the Chief asked.

I shook my head. "Wrong place. Wrong time. We did what we could."

Mason and I left by the front door. It was getting hot, so we hurried across the pebble- covered driveway to our air-conditioned house.

"Why did you tell him that, Mason?"

Mason grinned at me. "It's true, but mostly I like to say 'my wife.' I've waited a long time to say that."

He wrapped his arms around me and lifted me off my feet. "I could carry you over the threshold."

"Not necessary. This isn't our house, remember?" We walked inside and sat on the couch to cool off. "Let's give the kids a call."

Lurleen answered. "Danny has taken the kids to Apalachicola. I heard about the news on the TV—Brad Arborough was found dead on the beach. The news said is wasn't clear if foul play was involved. What's the scoop?"

"Sounds like you know as much as we do," I said. "First they found his boat. Later they found his body. Lisa said he loved that boat, so the speculation is he got drunk and fell overboard. The coroner's report is coming in a couple of days."

"Whoa," Lurleen said.

"There's more," I said. "Mason found an emerald and diamond bracelet under a pillow in the master bedroom, and he and the Chief discovered what looked like old blood on a sheet."

"*Ooh, la la,*" Lurleen said. Lurleen switched into her unique version of French whenever she got excited. I could hear her mind working.

"Who slept in that room?" I asked.

"No one," Lurleen said. "It was a funny thing. It's the best room in the house, and no one felt like claiming it. We settled in the other bedrooms. So does anyone think Brad might have been killed in that room?"

"No, not enough blood, but he was staying in the house before we all came. Lisa said he probably entertained women there and that he liked to play a little rough. The blood might have been from scratches, that sort of thing. He'd give his girlfriends expensive gifts—like the bracelet—when he was ready to break up with them according to Lisa. She said he used her money to buy them."

"Wow. What are you and Mason going to do about it?"

"It?" I asked.

"The murder."

"We don't know it's murder yet," I said.

"I do," Lurleen said."Brad made people furious. He and Lisa were in the midst of a terrible divorce. Brad got into trouble every time he went into a bar on Cape Azul, which was all the time. Most of the men around here worried he was fooling around with their wives. He was murdered, Ditie, no doubt about it."

"I don't know how you know all that, but I hope you're wrong," I said. "Mason and I want a quiet honeymoon."

"You can have that, but I'd like to visit you and the scene of the crime," Lurleen said.

"Lurleen, it isn't the scene of the crime, and hopefully there was no crime."

"*Mon dieu*—you know that's not true. Honestly, Ditie, doesn't it get your juices going just a bit?"

"My juices are going for entirely different reasons. I'm finally married to the man I love."

"I know, I know, but that can keep for a while. I'll be over in ten minutes."

I got off the phone and told Mason how Lurleen was reacting to the news.

He smiled. "You didn't expect her to leave it alone, did you?"

I shook my head. "No, not really."

"I have my doubts that you can leave it alone either," Mason said. This time he wasn't smiling.

"And what about you?"

"Ditie, I'm on vacation with my new wife. This is none of our affair." He pulled me into a tight embrace. "Ten minutes isn't really enough time to do anything."

"We could do a little."

We kissed like we were teenagers. Lurleen didn't knock and we hadn't locked the door.

"Oops," she said when she saw us. "Well, that can wait. Tell me all you know, and I'll tell you what I've heard."

9

"Let me see the bracelet," she said as soon as she got settled on the couch. "I can tell you if it's real or not."

"The police have it," Mason said.

I showed her the photo I'd taken.

"It looked real to me," I said.

"Yes, but you can be fooled by fake jewelry. Remember the necklace Savannah Evans wore in 'The Case of the Purloined Necklace?' "

"What are you doing, Lurleen?" I asked. 'You're naming each of our old cases like Dr. Watson did for Sherlock Holmes?"

"Why not? They're every bit as good as his adventures. Even *I* was fooled by that necklace, but that was because I didn't get a chance to see it up close. It's really hard to tell much of anything from a photograph. If it is real, it's worth a lot of money. When can I see it?"

"Sorry, Lurleen," I said. "You won't get to see this one up close, but Lisa Arborough seemed convinced it was real. According to Lisa, he used her money to buy it."

"Wait a sec," Lurleen said. "Lisa is the one who has the money? I thought it was Brad."

"There seems to be some difference of opinion about that," I said. "As it stands now, it looks as if Lisa made the big bucks. Brad had a series of young girlfriends, and Lisa put up with it. She says it was Brad who finally filed for divorce, and I'm not sure she wanted that divorce to go through—despite the fact she claims she was paying for Brad's bad behavior. It doesn't make much sense."

"No, it doesn't," Lurleen said. "She's an attractive woman —a lot of men would be interested in her—and here she is, putting up with a man running around on her. Maybe she finally got fed up with that and did him in."

"Maybe," I said, "but she seemed genuinely upset about his death."

Mason hadn't said a word. He stood, paced a bit and stared out the window at the ocean.

"What is it, Mason?" I asked.

"I'm on my honeymoon," he said. "I thought I'd have you to myself for a few days, Ditie. I don't mean entirely to myself, but I do mean away from worries about work or unsolved murders. I don't want a busman's holiday."

"Such a cute phrase, Mason," Lurleen said and patted the seat beside her. "Didn't that come from a book by Dorothy Sayers: *Busman's Honeymoon*? I still love those mysteries. Relax, sit down, no one's asking you to get involved."

Mason remained standing. "Come on, Lurleen. You get involved, you get Ditie involved, and I'm involved! I just want some quiet time with my wife at the ocean watching out for dolphins and eating too much seafood."

Mason looked forlorn—that was the only accurate description. The corners of his mouth drooped. He'd even raised his voice to Lurleen—something I'd never seen him do before.

For a moment the three of us sat in silence.

I stood, moved around Lurleen and put my hand on his

arm. "I'm sorry, honey. I want exactly what you want." I turned to Lurleen. "This is one mystery we can't get involved with. For all we know, Brad fell off his boat. And if he didn't, his death is in the hands of the police. We aren't needed here. You have the kids to look after, and I have Mason."

Lurleen nodded slowly. "I understand. I'm sorry I upset you, Mason. I just had news of my own to share but it can keep."

Lurleen was sneaky. She knew precisely the way to reach me—tell me she had information and then refuse to share it.

Mason sighed. "What is it you know? Have you shared it with the police?'

"I wouldn't know who to call," Lurleen said, "but you would. You could tell me if it's information the police should hear about."

"What does Danny say?" Mason asked. Danny was an ex-cop, now PI. He'd know what to do with Lurleen's information.

This time it was Lurleen who pouted, much more dramatically than Mason had. "He doesn't want to hear it. He says he has his hands full with four kids to entertain, two of them young teenagers. I told him teenage girls don't want to be entertained—they want to be left alone."

"So you haven't told him what you think you know?" Mason asked. I recognized the tone in Mason's voice. It was the same tone he used when he told me about witnesses who refused to give an honest account of a crime. Exasperation—that was the tone. I'd never seen him talk that way to Lurleen.

But then again, Mason had been waiting a long time for us to get married. He'd waited through murder investigations, waited for me to adopt the kids, waited for us to plan the perfect wedding. He'd been a very patient man, and now all he wanted was to be with me.

"Lurleen," I said. "I don't want to hear what you've discovered." That wasn't entirely true, but it was close enough.

"Here's the Chief's card. His name is Scott Huxley, and he seems like a competent officer. Tell him whatever you know, and let him handle it."

Lurleen stood up and looked down at us. "I would do that," Lurleen said, "but there's a chance you two could be in danger."

10

"In danger?" Mason and I said together.

"You come here," Mason said, "and leave this piece of information out of the conversation until the moment we ask you to leave." Now there was no mistaking Mason's tone. He was angry.

I wasn't sure how seriously to take Lurleen's warning. She could sometimes exaggerate to make a point.

"Sit down," Mason said, "and tell us what you're talking about."

What he meant was that whatever Lurleen was about to tell us, it better be good and it better be accurate.

"Anyone else feel a chill in the air?" Lurleen asked. "Maybe I should turn down the air conditioning."

"Sit Lurleen," I said, handing her a throw in case she actually was cold. "First, tell us if the kids are in any danger."

"No, no," she said. "It's your location that's putting you in danger. You are next door to the big house in which Brad Arborough was killed, in which he had affairs. The Arboroughs own both these houses. And it seems they both had affairs."

"And you know this how?" Mason asked Lurleen.

"I get out and about. I talk to people. You know how I am, Mason. You sound like you're accusing me of something."

Mason sighed. "There is no evidence Brad Arborough was killed in the big house."

"Well, his body turned up next to it," Lurleen said. She sounded a little defensive."And if his body turned up nearby, then someone is making a statement."

"Or maybe," Mason said, "he fell off his boat, drunk as people speculate, drowned and the tide brought his body in where it was found."

Lurleen nodded. "Maybe. But you should hear all the rumors."

"No," Mason said. "The Chief of Police Scott Huxley should hear all the rumors. And I suspect he already has."

"I'm sure he has," Lurleen said, "but you don't seem to realize Chief Huxley may be part of the problem. People say he and Lisa Arborough have been in a relationship for years."

"For years?" I asked.

"What people?" Mason asked.

Lurleen turned first to Mason. "Everyone on the Cape knows Scott Huxley and Lisa Arborough are an item."

Then she turned to me. "When Lisa and Brad began taking their vacations separately, they were free to do what they wanted with whomever they wanted. Kind of like an open marriage."

"If that's true," Mason asked, "how does that put us in danger?"

'It's not so much Brad and Lisa who put you in danger. It's more about who might have killed Brad—and my sources say he *was* murdered. If the murderer thinks you know too much, you could be in danger."

"What sources?" Mason asked, but he didn't wait for Lurleen to respond. "Look, Lurleen, we know nothing, and we plan to keep it that way. You are making a very good case for staying out of this investigation, and that's what we intend to do."

"That's what we want *you* to do, Lurleen," I said. "You have the children to look after."

"It's not that simple. People on the Cape know exactly who you are. They know you are a big city detective, Mason, so they wonder why you're here."

"My wedding isn't sufficient?"

"It would have been until the murder happened," Lurleen said. "And they know about you, Ditie."

"They know I'm a pediatrician, also here for my wedding?"

"They know about the other murder cases you've solved."

"How would they know that, Lurleen?" I asked.

"You have to give information to get information," Lurleen said.

"Lurleen," I said, "all I want you to do is take care of the kids. Danny can't be happy about this."

"Danny took the kids on a boating trip up the estuary from Apalachicola. He's gone today. All day. I've been left to my own devices."

Never a good situation, I thought.

"I drove him and the kids to Apalachicola this morning," Lurleen said. "I'll pick him up this evening. In the meantime, I've talked to a lot of the shopkeepers including a jeweler who has made jewelry for Brad."

"Including an emerald and diamond bracelet?" I asked.

"Not the one you found," Lurleen said. "Of course, I didn't know about that until you just told me. But I heard about his parting gifts. Mary owns a store called Lost Horizons, and she said she'd made a lot of pieces for Brad. Long ago she made a bracelet with emerald and diamonds Brad gave her, but she only works in silver, I think. So that can't be the one you found.

"Recently, she made a piece for Scott Huxley. She called it a dolphin pin—silver with sapphires."

"I saw Lisa wearing that pin," I said without thinking.

"Exactly," Lurleen said. "What more proof do you need that they are an item?"

"This has gone far enough," Mason said. "Every bit of information you've told us needs to go to the police." He looked at me. "We are staying out of this."

"But Mason," Lurleen said, "don't you understand—Scott Huxley might be in the middle of all of this? I can't take this information to him."

Mason sighed. "You document everything you've heard. You take it to the police and give copies to his deputy. You keep a copy."

"When are Danny and the kids getting back?" I asked.

"Late, I think. Danny's going to call me. If it's too late, he'll get rooms at the Apalachicola Inn, and I'll pick them up tomorrow morning."

"I think we need to take some time to sort this out," I said. "We'll come over tomorrow. In the meantime, Lurleen, please don't talk to another living soul about the murder— if it is murder."

"I second that," Mason said.

"I was only trying to keep you informed," Lurleen said. "You're close to the action whether you like it or not."

"Ditie and I will talk this over," Mason said. "If we need to move out of this house, we'll do that. No more conversations with anyone about the death of Brad Arborough, is that clear, Lurleen?"

"Clear," Lurleen said.

She left without a goodbye.

I hated it when two people I loved were in the middle of a serious argument, but Mason was right. We wouldn't be in any danger if Lurleen hadn't bragged about other murder cases we'd solved or about the fact Mason was an Atlanta detective.

If Brad was murdered, his killer might be very nervous about having us around.

Mason closed the door after Lurleen left and sat back down on the sofa beside me. "We tried to do this so carefully," he said. "I have to admit I'm frustrated."

"I know. Me too," I said. "Brad Arborough got himself killed one way or another, and now everything seems to be spiraling out of control. Should we pack up and head back to Atlanta?"

"The kids will be so disappointed," Mason said. "Worse than I am, maybe. We're married now, and that's all I've wanted for this last year. I don't need the perfect honeymoon. But the kids—they've been so excited about the trip."

"You're right, Mason, and they aren't in danger. I'm not convinced we are either. Let's put everything on hold for now. Maybe Brad really did fall off that boat when he was drunk. Can we find out what the coroner says?"

"I think I can make that happen," Mason said, "but it won't be for a day or two."

"Okay. Sounds as if this afternoon and evening belong to us. Let's make the most of it."

11

We knew the kids were safe with Danny and Lurleen. For the first time in a long time, I was glad they weren't near us. I missed them, but I knew they were having a wonderful time. Mason and I were determined to do the same.

"Wine on the deck?" Mason asked.

I nodded.

"Would you like dinner out somewhere?" he asked.

"I'd like to stay in," I said. "We could get some fresh shrimp and have a simple meal here."

"That sounds perfect with one small exception. It's still pretty hot out there. Maybe we could spend a little time upstairs and let things cool off outside."

"I'm for that."

We spent the next hour just the two of us together. No one pounded on our door. No one tried to reach us by phone. It was a miracle and we enjoyed every minute of it.

Then we put on bathing suits and headed for an afternoon swim. The water in the gulf always soothed me. This afternoon was no exception. The sea was calm, the water clear enough to spot any string rays that might be passing by. We wore our water shoes and shuffled our way into the ocean. Mason told me he loved the beach and the ocean, but his former wife did not. She found that kind of vacation too quiet for her tastes, so most of their vacations had been to livelier places.

"This water is so warm," Mason said.

"Exactly the way I like it."

We spotted one or two sting rays, but they weren't spawning like they'd been on an earlier vacation when the water was thick with them.

"If you shuffle through the sand, you'll give them enough warning," I said. "They don't really want to sting us, and if they do, we'll head for a hot bath."

"And the jellyfish?" Mason asked.

"I haven't seen any on the beach, but if we get stung, we'll pour vinegar over the wound, pull out any tentacles we can find and then take a hot bath. We've been really fortunate so far—no black flies, no jellyfish, only a few sting rays. Mostly sunny days. Only fly in the ointment, so to speak, is the unfortunate death of Brad Arborough."

Mason put a salty finger to my lips. "We'll discuss that when and if we have to. Right now, I'm a very happy man. The wedding was wonderful. We're married. The kids are having a great time and we get to be alone together."

"Not another word about dead bodies, I promise."

We stayed in the water as long as we wanted. Then we showered, dressed, bought shrimp, and settled on the deck with wine and cheese. I called the kids. They were excited about their boat trip with Danny and planned to stay in Apalachicola for the night.

"Mama," Lucie said. "Did you know what an old river town Apalachicola is? Did you know the Chattahoochee flows from Atlanta and joins the Flint River to become the Apalachicola River and that the city is built on the Apalachicola Bay? Do you know how small the town is? Only about three thousand residents. Do you know about Mr. Gorrie who invented the ice machine? He has his own museum here. There is still so much to know about this town, Mama. I'm keeping a journal of everything we see and do—I'll show you tomorrow. Are you and Mason having fun?"

"Yes!"

Jason was just as excited. "Noah hasn't stayed overnight in a hotel away from his family ever."

"You haven't either, have you Jason?"

"No. Danny says we can eat popcorn and watch movies tonight. Is that okay with you, Mommy?"

"More than okay—you're on vacation. Uncle Mason and I might do the same thing."

Jason was quiet for a moment.

"What is it, Jason?" I asked.

"Mommy, is Uncle Mason going to adopt us, like you did?"

"Of course. We'll have another party when that happens. You'll have a second dad—is that okay with you?"

"Will I still call him Uncle Mason?"

"You can call him whatever you like," I said. "Mason, Uncle Mason, Dad—whatever, Jason. He will think of you as his son—he already does. You can try out some different names the way Lucie did with me. You'll still have Charlie Flack as your first dad, but Mason will be your second one, and he'll always be there for you."

"Danny's calling me, Mommy. I have to go."

We hung up the phone, and I told Mason about our conversation.

"Do you think he's worried about having me as his father?" Mason asked.

I shook my head no. "So much has happened in his young life," I said. "I think he simply wants security without knowing how to talk about it. He's confused about having two fathers—any kid would be—but he'll see how reliable you are. He already has some idea about how much you love him. It will all work out."

I gave Mason a hug. Then I fixed a simple shrimp dinner —shrimp Alfredo with rice and snow peas. Mason opened another bottle of champagne.

It was a perfect evening but a short one. That's what the sand and water and summer heat do to a person. We watched the sun set and we were in bed and asleep by ten.

At eleven, Mason got a call on his cell. He rolled away from me to check the number.

"Don't know the number," he said. "If it's important, they'll leave a message."

Seconds after the initial call, we both heard the ping of a text message. It was from Lisa Arborough.

"Lisa here. Sorry to bother you at night, but I thought you should know. Scott heard from the coroner. Brad was dead before he hit the water. No water in his lungs and a blow to the back of the head. Someone murdered him. If you want to leave the Cape, let me know. I'll refund your money. Also don't be alarmed about what's going on next door. I'm helping Wayne move in—Scott gave us the okay to use the house."

I looked over Mason's shoulder as he texted back. "We'll discuss this and make a decision in the morning. Thanks for letting us know."

"Ugh," I said. "Now what?"

"Nothing for tonight. You think you can sleep?" Mason asked. "I'm not sure I can."

I turned on the light. "How about popcorn and a movie—that's what the kids are doing."

"I'm in," Mason said. "I'll pop the popcorn and you find the movie."

Mason and I shared a love of old movies, and I managed to find a Miss Marple mystery starring Margaret Rutherford in a collection of DVDs near the TV. " 'Murder Ahoy' sound okay to you?"

"It's about death on the high seas," Mason said. "Too close to home?"

"Murder from a snuff box wasn't it?" I said. "Smugglers, financial shenanigans."

"You sound as if you watched it yesterday," Mason said.

"No, but it might shed some light on what's happening here. You think Brad was into some kind of smuggling—drugs, something like that?"

"I doubt it. It sounds as if Brad Arborough was broke and got his money from Lisa."

"Lisa might be a prime suspect," I said, "especially if Brad was using her money on other women."

"Did you forget we're not into this murder? We're only into each other."

I snuggled beside him in bed. "I didn't forget—just testing your resolve."

"It's not *my* resolve I'm worried about," Mason said. "I'm not positive Lurleen can let this go, and I'm not so sure you can either. You two get curious about an unexplained event and you don't stop until you figure it out. All I want is a honeymoon with my new wife."

"You shall have that, sweetie," I said.

We didn't see a lot of the movie, which didn't matter. Both of us had seen it before more than once.

"I love you, Mason," I said. "I love the way you are with my friends, with the kids."

"And I love how welcoming you are to my sons and their families," he said. "I'm sorry about the argument with Lurleen. I got worried that somehow we wouldn't have this time together. I'm better now."

"Lurleen can't help herself when a murder falls into her lap, but the children are her first priority. She'll never put them in harm's way."

That seemed to be enough reassurance on both sides to let

us fall asleep. It wasn't a banging that woke us up the next morning. It was a phone call.

I recognized the number—it was Lisa's.

"I hope I'm not disturbing you too early in the morning. Wayne bought doughnuts from the pop-up vendor near us, and we can't possibly eat them all. We'd love to share them and see what you've decided about going or staying."

"Give us fifteen minutes," I said, "and I'll put on some coffee."

Mason jumped in the shower and I followed. We were dressed and downstairs ten minutes later.

Wayne and Lisa were at our door a few minutes after that. They looked like a couple, and I wondered if they'd been enjoying each other's company in the same way Mason and I had.

"Inside or out?" I asked.

Lisa looked at her watch. "It's only ten—it should still be cool enough outside. Weather forecast says we'll get a break from the heat today, and this house has a great overhang from the deck. It's why I bought it."

"Did you buy this house first and then the one next door?" I asked.

"Other way around. Brad and I had plans to make the house next door a place where all our friends could gather. That was shortly after we married when we still had mutual friends."

We settled around an outdoor table with our coffee and doughnuts.

"When was that?" I asked.

Mason gave me a look, which I ignored. I wasn't going to get into a hornet's nest with Lisa, but it wouldn't hurt to get a few facts straight.

She seemed eager to talk. "I'd been coming to the Cape several years before I met Brad. We met in Tallahassee at a Florida real estate convention, but we didn't stay here together until after our marriage ten years ago. The house next door came up for sale while we were here. I told Brad what a good deal we could get on it, and he agreed. At that point he had some of his own money—or at least I thought he had—so we put the house in both our names. I ended up with all the mortgage payments, and that's one reason I was determined not to let Brad get the big house in the divorce."

She paused. I thought she might be feeling emotional about Brad, but in fact she only wanted another doughnut.

"These are so good," she said. "I hope you don't think I'm being crass about Brad's death. I'm not, honestly, but I can't pretend feelings I don't have. Brad did everything he could to kill our relationship, and he finally succeeded. If you want to find sympathetic folks, you should look to Brad's guy friends. Some of them did as much partying and sleeping around as he did. They'll probably tell you what a good guy he was."

Wayne made a face. "I'm not sure they'd do that anymore. Brad tried to borrow from every one of us."

"You told me you were here," I said, "because he promised to pay you back the money you loaned him. You believed him?" I asked.

"Shouldn't have," Wayne said. "Actually, Lisa is more the reason I came down. I wanted to offer her support during the divorce—I knew Brad would try to make it hard on her."

"I thought you didn't know about the divorce until you got here," I said.

Wayne was silent for a moment. "I called Lisa to tell her I was coming down—that's how I heard about the possible divorce."

Lisa took Wayne's hand and squeezed it.

"I remained so naive about Brad," she said. "I imagined we might still reconcile."

Lisa didn't strike me as the naive type, although I did know love could do funny things to people.

"I went to the first mediation session hoping for that."

"Why?" I asked. The words were out of my mouth before I thought about them. Mason shot me another look that said I was getting in too deep.

"You don't have to answer that," I said. "It's that you sounded so done with him just now, and you seem to have a lot of admirers here hoping you'll move on."

I didn't mention Lurleen's rumor that Lisa and Scott Huxley were sleeping together. I wasn't sure how Wayne might react to that.

"I was surprised when you said he initiated the divorce," I said. "I assumed you'd done it. If I'm asking too many personal questions, I apologize—we can switch topics."

"It's all right," Lisa said. "Your questions make sense, and it's nice to talk to someone who's not directly involved in the whole mess. Brad kept promising me he could change, and I kept believing him. I even wondered if the bracelet might have been meant for me—our tenth anniversary would have been tomorrow."

She shook her head and stared at the large diamond she still wore on her hand. "Of course, I realized after the first mediation session, Brad didn't intend to reconcile. He wanted to get as much as he could from me."

"But he was the one having the affairs," Wayne said. "I never understood why you owed him anything."

"It's a little more complicated than that, Wayne."

Lisa sighed and smoothed down her platinum hair. Not that a hair was out of place. She was perfectly put together the way I'd always seen her with a colorful summer outfit and a tan that might have been real or a spray on.

"There was a time I was down on my luck," she said, "right before I met Brad. The housing market had tanked. I had debts from school. A lot was going on. Brad came from a wealthy family, and he agreed to help me out—so you see some of my start-up money came from Brad. He wanted to get that money back in the divorce settlement. He said he had plans about how he wanted to use it."

Wayne looked surprised. "I never understood why he didn't just hit up his dad for more money—why he needed to borrow it from me."

"His father disowned him five years ago—that's why," Lisa said. "The old man had had enough, and I don't think the Senior Arboroughs had much money left after Brad's divorce from Shannon."

Mason stood. He'd also clearly had enough. "I'll be happy to make another pot of coffee," he said.

Lisa looked at her watch. "Oh my gosh, I'm sorry. We have to go. I have an appointment with the Chief in half an hour, and Wayne . . . Wayne has to finish settling in."

"Yes, right," Wayne said. "I may be staying for a while."

"So, did you two make a decision about leaving?" Lisa asked.

I looked at Mason. "We haven't really talked about it yet," I said, "but I know there's nothing else to rent on the Cape. Our kids are having a great time, and I don't want to disrupt them—unless we have to."

Mason nodded. "For now, we'd like to stay through our rental agreement. We'll let you know if that changes."

Lisa didn't look particularly happy with our decision. "As you wish. If I find another place as nice as this, I'll call you."

"Thanks," I said.

Mason showed Lisa and Wayne out the door.

"Lisa didn't seem to like the fact we were staying on," Mason said.

"No, she didn't."

"Makes you wonder," Mason said.

12

"I knew it," I said. "You can take a detective out of the city, but you can't bury his curiosity, even for a week."

"Wanna bet?" Mason took my hand and led me upstairs. "No more talk of murders or motives," he said. "No more talk at all."

We resurfaced at noon, and I called Lurleen to see if we could meet the kids for lunch. It had been two days since I'd seen them, and it felt like a lifetime.

"Come here," Lurleen said. "Danny will fix us some lunch. I have a lot to tell you."

Uh oh.

I didn't tell Mason what Lurleen said.

Lurleen, Danny and the kids were staying in the condo we'd rented for the past two years—half a mile south from where we were. It sat back from the beach protected by the dunes and some shrubbery, but we could still see and hear the ocean from the deck.

The kids scrambled all over us when we arrived. It felt like the welcome Hermione gave us when we returned home after a brief shopping trip. While Hermione had to settle for jiggling her entire body and sloppy licks on our hands, the kids gave us hugs and then told us about all their adventures from the day before.

The sky was overcast with dark clouds hovering above our condo, so we ate lunch inside. Then we stayed inside when the downpour erupted and didn't let up.

No one spoke of murder except when we played Clue. The kids wandered off to their rooms around four and that left Lurleen, Danny, Mason and me to our own devices.

Lurleen was clearly itching to catch us alone.

"I'm not sure how soundproof these walls are," Mason said before Lurleen started talking.

"I'll keep it down," she promised. "This is more of an FYI moment anyway." Lurleen flicked back her hair, and her hazel eyes sparkled with excitement.

"I think we better hear her out, Mason," I said, "before she pops."

"Please do," Danny said. "Otherwise, she'll keep me up all night with her ideas about the Arborough murder."

"I'm right here," Lurleen said. "I'm not hysterical and I'm not bringing you into the case."

That worried me. "I hope *you're* not into the case, Lurleen. Your job is to keep the kids safe and entertained."

"I know that." She started to pout, her lower lip protruding, and then she apparently thought better of it. "There's no time for getting upset. The kids will want a snack any minute now. Here's what I know. If you don't want to hear it, Mason, you can go upstairs and check on the kids or start dinner—we're having tacos."

"I'm sorry I was harsh with you, Lurleen," Mason said. "It's just that I've been waiting a long time for a honeymoon with Ditie, and I didn't want anything to interrupt that."

"I get it. You're forgiven." She paused probably to make sure she had our complete attention. "I've become friends with the Chief of Police, Scott Huxley, and he can get very talkative over a couple of beers."

"Friends?" I asked. "When did you have time to see Scott Huxley?"

"Danny was gone with the kids all day yesterday and last night, so I had plenty of time."

Lurleen with time on her hands was rarely a good thing. She'd find a way to fill it, and if there happened to be a murder in the vicinity, she'd be busy gathering information.

"You don't need to give me that look, Ditie. I should have said Danny and I have become friends with Scott. Turns out Scott is into every private detective show he can find on TV and wants to know Danny's take on how realistic they are."

"The Chief asked for my help with Brad's murder investigation," Danny said.

"You agreed to that?" Mason asked.

"I said I'd consider it. The case is becoming pretty interesting, Mason," Danny said. "I was busy with the kids in Apalachicola for the last two days, so Lurleen is the one with the information. Scott's going to call this evening with an update."

"Scott," Mason said.

"It's really informal around here, Mason," Danny said. "Everyone's on a first name basis."

"And the kids?" I asked. "Have you managed to keep them out of this?"

"I think so," Lurleen said. "With Lucie, you never know what she's hearing or what she's writing down in her notebook. But the kids are in no danger, I can promise you that. It feels like this murder was a long time coming. Maybe a one-off."

"You can't know that, Lurleen," Mason said. "You go digging around and scare a murderer, and there's nothing to keep him from acting again."

"Or her," Lurleen said. "Anyway, I'm not digging. I'm leaving it to Scott. If he asks for my advice, I'll give it."

"And you, Danny?" Mason asked. "Are you fine with all

this? How involved are you. Did you agree to help Scott Huxley?"

"Not yet, so we're not involved," Danny said, "and the kids are not involved. Noah's mom heard about the murder and she's worried, naturally."

"How did she hear about it?" I asked.

"Maybe I said something when she called," Lurleen said. "You didn't plan to keep it a secret from her, did you Ditie?"

"No. I was waiting until Mason and I made a final decision about going home early."

"Now, I don't think you have to do that," Lurleen said. "Noah's mom decided she'd come and stay a day or two with us and then take the kids back with her. Has she called you about that yet?"

"No," I said.

"She will," Lurleen said. "Her older boys are at camp for another couple of weeks, so she's happy to entertain all four of the kids—make it like their own summer camp in Atlanta."

"That's probably a good idea," I said, "but I'll miss them if we do stay on another week."

"Let's hear what Lurleen has to say before we decide about that," Mason said.

"*Bien.* Is it my turn at last?"

"You have the stage, Lurleen," I said.

"Okay, then I want to start at the beginning. Well, not at the very beginning. The real beginning to all of this lies in Apalachicola—that's what I think, anyway. I'll get to that later."

"Fine," I said. "Start with the chief complaint, the history of present illness—that's what I do as a doctor."

Lurleen nodded and smiled. "Okay, chief complaint—a guy is dead . . . murdered. As to the history of the present illness, which starts before his death I guess, is that Brad made a lot of people furious. Brad Arborough had a really bad reputation. He threw money around when he had it and scrounged from his friends when he didn't. He was a blowhard and a bully, worse when he was drinking. But he did apparently have one secret weapon."

Lurleen stopped talking to see if we would take the bait. We did.

"Go on, Lurleen," I said. "We're listening. What was his secret weapon?"

"Brad had this amazing ability to seduce people. I don't mean just women, although he was good at that. Brad could seduce almost anyone initially. He could make you believe everything that came out of his mouth. Did I mention Brad was a serial liar? It was never clear if he actually believed his own lies, but it was clear nothing he said proved true."

"How do you know this, Lurleen?" I asked.

"From Scott. Then from your neighbor, Wayne. He thinks Lisa believed some of Brad's lies—like the idea he wanted her back and would change his behavior."

Danny leaned into the conversation. "The one thing Lurleen has left out is that Brad never thought he had to change anything about his behavior. He never admitted he lied, and he never admitted he'd done anything wrong. The problem was always the other guy."

"I'd like to hear the details of how he worked his magic," Mason said. "Start with the Chief of Police, Scott Huxley. He's the guy who should know better."

Lurleen nodded. "That's what makes this all so interesting. Even Scott was duped, and he's had his eye on Lisa for years. Do you want to tell it, Danny? How does a guy like Scott Huxley believe in a guy like Brad Arborough?"

"It's your story, Lurleen," Danny said, "you tell it. I'll listen out for the kids and add my two cents if I need to."

"Okay, *chèri*." Lurleen settled back onto the couch. "I could do this better with a glass of wine."

"At your service," Danny said. "Anyone else?"

I nodded. I was sure I could hear better with a glass of wine. Mason and Danny refrained.

"This is where Apalachicola comes in," Lurleen said. "Scott met Brad and Lisa in a bar in Apalachicola several years ago. Lisa was checking out the area for real estate opportunities and thought she might open an office in town. She saw the area as low-hanging fruit—meaning she expected the town to grow exponentially in the next few years."

Lurleen paused to sip her wine.

"Scott wasn't sure if that idea came from Brad," she said. "All he knew was that Lisa was on board with expanding her real estate business into Apalachicola. Brad seemed to have his own ideas about new businesses that would flourish in the area. He got Scott on board. Scott, who'd lived in the area for thirty years, suddenly saw it as a gold mine. He invested his retirement savings into Brad's vision and lost every penny. He lost his wife as well—their divorce was final two years ago."

"I assume Scott was bitter," I said. "Was he bitter enough to kill the guy?"

"That was the odd thing," Lurleen said. "He didn't sound bitter when he told me the story. He still sounded—like he believed in the guy. It was really strange, *très, très bizarre*. You spoke with him, Danny, what did you think?"

Danny sat beside Lurleen and nodded. "He acted as if he was upset about Brad's death because he still thought Brad could have pulled off his plans. He said he thought Brad would make a killing in Apalachicola eventually, and it was too bad he got himself killed first."

"This was the same guy who had feelings for Lisa," I said. "Do I have that right?"

"You do, Ditie," Lurleen said, "but I think Scott is more of a flirt than an action guy. I got the feeling he admired Brad's ability with women. I also got the feeling he was devastated by his own divorce. He sounded madder at his wife for leaving him than at Brad for losing his money."

"Even though Scott lost all his retirement savings?" Mason asked.

"Yep," Danny said. "Scott claimed his wife needed to be patient, to have faith in Brad's ideas. It was pretty remarkable."

"Danny, do you think Scott was putting on a show for you?" I asked.

"Maybe. I barely know the guy, and he'd want to look good for me, I suspect."

"What did you think, Lurleen?" I asked. She was generally a good judge of character.

"I wondered about that," Lurleen said. "Brad had caused Scott nothing but grief, so I wondered if what he was saying wasn't really how he felt."

"Are you thinking he might be involved with the murder?" I asked.

Lurleen shrugged. "He worked pretty hard to convince us he was upset about Brad's death. And then he asked Danny to help with the investigation—said he needed an objective eye. Why would he do that if he were involved?"

"Good question," I said. "What about the women Brad seduced? I mean the women he literally seduced? Are any of them suspects in the murder?"

"That I'm not privy to," Danny said. "Scott wouldn't give out names."

"But," Lurleen said, "if you ask around, you get a sense of how women felt about him. Men, too. Brad Arborough stirred up a lot of emotions. People loved or hated him, not much in between."

"Scott has a list of suspects?" Mason asked.

"It might be easier to say who's not on the list," Danny said, "but what's most curious is how the guy died."

That got our attention.

"You heard the coroner's report," Danny said, "no water in his lungs, so he didn't drown—he was dead before he hit the water. Blunt trauma to the back of his head, but no blood on his boat. And no fingerprints on the boat except for Brad's and the kids who found the boat adrift. If he was murdered on that boat, someone did a very good job of cleaning up the blood."

"Yes," Lurleen said. "It's like a locked room mystery—only this one is on the open seas."

13

"No signs of a fight on board his boat?" Mason asked.

"Nothing," Danny said. "No sign of anything amiss."

"What about the blood on the sheet in the big house?" Lurleen asked. "Could he have been killed there?"

"There wasn't enough blood to cause a death, but I suppose it's possible he was killed in the house," Mason said. "It doesn't seem likely, however. Someone would have had to get the body out of the house without being seen. If they did do that, they still had to get it into the water and then let it drift ashore—all of that with no one noticing. A lot of effort without much point to it."

He turned to Danny. "What about the kids who found the boat offshore? Has that been verified? And how far offshore was it?"

"Their story checks out," Danny said, "and it was only a hundred yards or so from where the boat was docked."

"That means if there had been a fight on board someone on land might have witnessed it," Mason said. "Did the coroner estimate a time of death?"

"Sometime two or three days ago," Lurleen said. "Maybe during your wedding. Maybe the night before."

"So that virtually rules out a murder in the house—everyone was here by then," I said. "I'm glad for that."

"Does the Chief have any theories about what happened?"

"Apparently not," Danny said. "If you want I can talk to him a little more about it."

"No," Mason said. "This is not our case."

"Got it," Danny said.

Lurleen was silent.

"Noah's mother will be down in two days," I said, "and we haven't seen much of the kids. That's where we should focus our energy. She can stay here with us if she likes—I'll call her and let her know I'm on board with her plans."

It didn't seem there was a lot more to say, which was a good thing as the kids tromped downstairs looking for something to eat.

I stood up. "Let me handle this, Lurleen."

I walked into the small kitchen with four hungry children behind me and searched the cabinets and refrigerator. "You have enough junk food to feed an army, Lurleen."

"Don't look at me—Danny bought all that stuff," Lurleen said.

"We're on vacation," Danny said. "Vacation equals junk food, and no one's complaining, Ditie."

That was true. Noah, Hannah, Lucie, and Jason grabbed their favorite snack and a soft drink from the refrigerator.

"No offense, Ditie," Lurleen said, "but you're really not needed around here right now. We've got this covered."

The kids settled on the couch, clicked on the TV and left me standing alone in the kitchen. I joined them on the sofa and watched a sci-fi movie of some kind. Mostly, I simply enjoyed sitting next to them, one arm around Lucie, the other around Jason. After an hour I started to get hungry.

"Dinner?" I asked Lurleen.

"You're welcome to stay, but we'll be eating late," Lurleen said. "The kids have barely finished their snacks."

"We're on ocean time," Danny added, "and that means we eat what we like when we want it."

"Maybe you and Mason should go home," Lurleen said, "and do whatever it is newlyweds like to do."

"Sounds good to me, Lurleen," Mason said.

The kids barely noticed when I stood up and we got ready to leave. Danny paused the movie. "Your folks are leaving," he announced.

"Bye, Mama," Lucie said giving me a perfunctory kiss. "We're having the best time."

"I can see that."

Jason let me hug him as did Noah and Hannah. Then it was back to the movie for them.

I was quiet on the road home.

"What's going on in that mind of yours, Ditie?" Mason asked.

"I guess I'm feeling a little dispensable. I mean—I knew when they got to be teenagers, they'd pull away from us, but they're not teenagers yet."

"Ditie, you have it all wrong," Mason said. "You're feeling dispensable because the kids are so secure. They can let you come and go because they know how loved they are."

"My wise Mason. I knew there was a perfectly good reason I married you. Let's get a drink somewhere and watch the sun set. Maybe we can catch that green light again."

"I'm not sure miracles happen on demand, but we can check it out."

We drove to the bar at the end of Cape Azul, the one located right in front of the sign and gate for the state park. A few cars were parked in front of the bar, but inside it wasn't crowded. It was dark and it took me a minute to adjust to the low lighting.

Mason asked if we could sit on the deck and the waitress was about to usher us outside when I noticed a familiar face.

"Mason, isn't that Lisa over there in the corner? Is that Wayne Baskin with her? I know she said they were friends, but they look like a lot more than friends to me."

Lisa was leaning across the table and Wayne was rubbing her arm. She appeared to be gazing into his eyes.

"Should we say hello?" I asked.

"Let's leave them in peace," Mason said, guiding me away from their table and onto the deck. "Who Lisa sees is her own business. I'm on my honeymoon, so I only have eyes for one person."

The waitress gave us a table with an unobstructed view of the water and the sinking sun. We looked over the list of summer cocktails.

"I'll bet these come with little umbrellas," I said to Mason. "I think I'll have white wine."

Mason ordered a Cape Azul Sunset. It arrived at our table a minute later—a blue concoction with a burst of orange liqueur in the center, and it did indeed come with an umbrella.

"I wanted to make sure you got this before the sun set," our waitress said.

"To us," Mason said, "and a hundred more sunsets with little umbrellas."

"I'll drink to that."

Mason moved beside me so we could watch the setting sun together. No green light appeared but there was plenty of magic. As the sky turned dark, sparkling lights appeared along the edge of the deck. We ordered appetizers of shrimp with avocado and cucumbers, coastal sliders, and crab rolls. This would be our dinner—more than we actually needed.

We left a little before ten. Lisa and Wayne were no longer at the corner table.

It gets very dark on Cape Azul with its two-lane roads and

rare street lamps. I let Mason focus on his driving, and that meant I had time to sort through what we knew about Brad Arborough and his death. He appeared to be a cheat in every meaning of the word, and yet people remained loyal to him. Lisa claimed she'd wanted a reconciliation, but here she was days after his death in the arms of another man.

I suppose that wasn't quite fair. They weren't in each other's arms, but they were awfully close to it. Lisa and Wayne were doing more than commiserating over the death of a man Lisa said she loved.

Was Lisa putting on an act when she said she'd hoped for a reconciliation? It was confusing. Brad had been a bully to Lurleen, but it seemed others still found him charming.

Someone had had enough—that was obvious. And the murder wasn't a simple act of passion. Someone carefully planned his murder or at least the disposal of his body. There was no evidence of foul play on his boat. Why, I wondered, did his body wash ashore so near the big house? Was that coincidence or a statement of some kind?

"You're thinking about the case, aren't you, Ditie?" Mason asked as he turned onto our drive. "I have to admit I am too. Why did his body turn up so close to the big house? The boat was nowhere near that house."

"My question exactly."

Our house was dark, but I saw lights on next door—inside and out.

"Doesn't Wayne know about the sea turtles coming ashore?" I asked. "Outside lights should be off and the blackout curtains drawn inside. I'm sure Lisa would have told him about that."

"I'll check," Mason said once he parked the car. "You get settled inside and wait for me.

He told me to lock myself inside until he returned. Then he took the gun he always carried and headed next door. He came back ten minutes later. "The front door wasn't locked," he said. "I went through the place. Someone had done that before me—drawers were pulled out, beds ripped apart. I didn't touch anything, but I did turn off the outside lights and pull the curtains. I need to call the Chief."

Mason called Danny and got Scott Huxley's number from him.

"I'll see if I can reach Lisa," I said.

Scott Huxley didn't answer, so Mason left a message. I was about to do the same when Lisa answered my cell. She sounded breathless.

"What is it?" she asked. I told her. She took her time responding.

"I'm with Wayne right now." I could hear a muffled conversation between the two of them. "He says he didn't leave the outside lights on and definitely closed the curtains. He isn't sure if he locked the door, but he's been here before—he knows about protecting the sea turtles."

"Where are you right now?" I asked.

"We're in a hotel in Apalachicola, having dinner, but we'll come home immediately—to the big house and then to you. Does it look like a break-in? We've had those before with kids looking for drugs and money. I'll call Scott. He lives on the Cape and can probably get there in ten minutes if I can reach him. Don't investigate anything on your own."

"We couldn't reach Scott," I said.

"I have his private number," she said. "I'll get him."

I hung up the phone. "Lisa thinks it's a break in by teenagers," I said. "She talks as if it's unrelated to Brad's death, never even mentioned that possibility. She also seems to be in a hotel with Wayne 'eating dinner.' Makes me wonder what's going on with those two. Wayne leaves his more than ample house to meet with Lisa in Apalachicola."

"Maybe they wanted a nice dinner," Mason said. "You can't get that on Cape Azul."

"It's eleven p.m., Mason. Who eats dinner at eleven at night?"

"I don't want you involved in this," Mason said.

"We are involved, Mason," I said, "like it or not. Lisa is calling Scott. She has his private number. We'll have to be available to talk to him when he comes."

We waited.

He arrived fifteen minutes later with his deputy. We watched from a window as he entered the house. Twenty minutes later, Mason's cell rang. He put it on speaker phone.

"I can't tell if anything's missing," Scott said, "I'll have to talk to Wayne Baskin about that, but someone did a good search of the house. Anyway, they won't have found anything of value."

"How can you say that for sure?" Mason asked.

"Because Wayne barely has two nickels to rub together—I know Brad screwed him over and I also know Wayne came to get his money back. Only a fool would leave anything of value in a rental."

"You seem to be forgetting the bracelet I found."

"Yeah, you're right, there was that."

"Why would someone turn on all the lights to search a house?" I asked.

"That's a question I can't answer," Scott said. "We've had a lot of break-ins lately. Teenagers looking for drugs. They aren't the careful types. Any sign someone tried to get in your house?"

"No," Mason said. "You don't think this might be related to Brad Arborough's murder? Maybe someone looking for that diamond and emerald bracelet? Someone who wanted you to think it was a bunch of rowdy kids looking for drugs or money?"

"Hmm," Scott said. "I thought you were keeping your hands out of this case."

"I am."

"Then why don't you let me do my work? I've been chief of police here for fifteen years."

"How many murders have you investigated?" Mason asked.

"I know what I'm doing and I don't need some big city detective telling me my business."

"Before you wanted my help, now you don't."

"I'm saying I can handle it."

Mason hung up the phone. I'd been standing close enough to Mason to hear every word.

"You seem upset," I said.

"Something doesn't smell right. Brad Arborough is murdered and no one thinks a break-in at the house he and his wife own might be related to his death or to the bracelet we found. Everyone seems a little too casual about investigating this murder. First, Scott Huxley is fine with my interest, and now suddenly he wants me to back off. Why the change of heart? I don't like any of it."

"I think you're right, Mason. Is there more we can do or should do?'

"We're moving out of this house tomorrow—that's one decision made."

14

We sat downstairs waiting for Lisa and Wayne to return. I was pretty sure they'd have questions for us, and Mason wanted to let her know we'd be leaving in the morning.

She knocked on our door around midnight.

"I talked to Scott," she said when we invited her in. "He seems to think it was a simple break-in, kids looking for drugs or money. Wayne is busy putting the house back together. All of this must be upsetting to you."

Mason nodded. "We've decided to move out in the morning."

"I completely understand," Lisa said. She sounded relieved. "Will you be going back to Atlanta? I'm so sorry if this has short-circuited your honeymoon."

"I'm not sure where we'll head next," Mason said. "Ditie and I haven't figured that out yet.

"You seem to be happy we're leaving," I said to Lisa.

"Only for your sake," Lisa said. "You shouldn't have to worry about anything on your honeymoon. Certainly not a horrible murder."

"It's nice you have such a close friend to support you right now," I said. "We saw you and Wayne together at the Blue Cafe, but we didn't want to interrupt."

Mason shot me a look.

"I'm not sure what you're implying, Dr. Brown," Lisa said. "You wouldn't have interrupted anything. Wayne is an old friend, and it's very nice to have him nearby. Now, if there's nothing more, I'll leave. There's no rush about your leaving in the morning. Take your time and text me when you head out. I'll send you a refund for your entire stay."

"Thank you," I said. "It's late. Do you have a friend you can stay with tonight?"

Lisa gave me a cold smile. "It won't be Wayne if that's what you're wondering. I've rented a house up Cape, ten minutes from here."

She left.

"What was that all about, Ditie?" Mason asked.

"I figured we were leaving, so she couldn't see me as a threat. And it might be the last opportunity I have to get information about what's really going on. Maybe she and Wayne didn't happen on each other here. Maybe they planned this reunion to get rid of Brad. With Brad out of the way, Lisa won't have to give him anything in a divorce settlement and maybe Wayne can get his money back. Maybe he and Lisa will ride off into the sunset together. I heard the two of them talking over the phone about his death—I wonder if that was a show for my benefit."

Mason shook his head. "You sound more like Lurleen than the level-headed woman I married. I'll chalk it up to too much excitement—what with marrying me."

"You are intoxicating," I said, "but you have to admit there's a lot that doesn't add up about Lisa. First, she says she wants a reconciliation; then she sounds as if she doesn't care that Brad is dead. I mean she tries to act as if she cares, but she has no emotion when she talks about Brad. People say she and Scott are an item and have been for years. And now we see her with Wayne. Who is this woman? That's all I'm asking. Some people say she was the moneymaker. Others say she was hanging around for Brad's money. Her story is full of contradictions."

Mason embraced me. "There is a lot we don't understand and may never understand. But what we really need to figure out is where we're going next. I still have vacation—you do too. I don't want to give up on that because we stumbled into a murder."

"Let's sleep on it," I said, "and see what we want to do in the morning."

We both slept soundly. That surprised me, but then the sound of the waves and the smell of the sea were the best sleep aides ever invented.

I called Lurleen in the morning.

She didn't sound happy. "You're leaving the prime spot for finding out what's really going on. That doesn't sound like you, Ditie."

"Lurleen, I'm on my honeymoon. I want to enjoy that. Mason and I both feel this murder investigation is sounding a little screwy, and we want to get as far away from it as we can."

"Screwy?" Lurleen asked. "What is it, *chèrie*? Scott has been very forthcoming with me. He says he has the case pretty much wrapped up. Brad was seen drinking in a bar with two strange men. They wanted to charter him for a trip around the tip of Cape Azul, offered him a lot of money to take them out. Then they disappeared. He doesn't know why they murdered him, but he's pretty sure they're the ones who must have done it."

"We heard none of that," I said, "but no matter what, Mason has had enough. After someone broke into the house next door, he's intent on getting away. I'm not sure where we're headed, but I know we're packing up and moving out."

"Come here, *chèrie.* The kids will love it. You can take the bedroom Danny and I have been using. We'll sleep on the couch. I'll check and see if anything is available down Cape—or maybe you want to get off Cape Azul entirely."

"I don't know. If we come, where will Noah's parents stay?"

"First off, it's only Noah's mom who's coming, and she said she had to delay her trip until Sunday. She can't stay over, so that's one problem solved. Come here first, and we'll sort out the rest."

I reported Lurleen's story to Mason and her invitation.

"I guess the strange men could explain why no one's linking the break-in to the murder," Mason said. "Two strangers—that's convenient."

I could practically hear Mason's mind working. He didn't want to be involved with this murder investigation, but he also couldn't stand to see a case mishandled. As for me, I always needed to know how the story ended.

"Maybe we take this one day at a time," I said. "Stay with Danny, Lurleen, and the kids for one night and then see if anything is available down Cape. It won't be as nice as this, but I won't mind that a bit."

"Agreed," Mason said.

The kids thought it was wonderful we were coming to stay with them, and in many ways it was. It was like camping out together. Early the next morning, Jason found me alone.

"I missed you, Mommy," he said. He wouldn't say that in public—Jason thought he was too big a boy to show his feelings around other people.

"I've missed you, too," I said.

"I think Noah misses his mom," Jason said. "He got upset when he heard she wasn't coming down until Sunday."

"Why don't we do something special—just the three of us today? Then it won't seem so long until his mom comes tomorrow."

I proposed a trip to a park, an open field really, at the

entrance to Cape Azul. Mason offered to take the girls anywhere they wanted to go.

"The bookstore in Apalachicola," Lucie said.

Hannah nodded. "Could we have lunch there?"

"Sure," Mason said. "I'm pretty sure Danny and Lurleen would enjoy a day to themselves. We'll bring dinner home with us."

Danny and Lurleen seemed pleased with the idea. "I'm happy to fix dinner," Danny said.

"Nope, you are off duty today," I said. "You can do whatever you want."

The boys and I left after breakfast. We found an open field in the park with a few kids playing on it. Noah had brought his soccer ball, and I'd brought a book I wanted to read. I sat in the shade and for an hour watched them play ball. Jason and Noah looked like a well-oiled machine, passing the ball back and forth. A couple of the other kids on the field joined in.

I'd brought snacks and water to share, and the weather cooperated to a point. It was overcast as if a storm was brewing but it didn't rain until late morning. At that point we piled into the car and headed home. Lurleen and Danny were gone I was glad to see.

While the boys showered and changed, I called Mason. "Why don't I fix the boys lunch, and you can stay in Apalachicola as long as you like?"

"The girls will love that. They've decided they need to go shopping. That means they need to look in every store in this town. I'll have some time on my own, and I've found us a place to stay unless it's too far from the beach for you. It's in the old Apalachicola Inn."

"I love being right on the water," I said, "but it would get us away from a murder that is none of our business."

"My thoughts exactly," Mason said. "Nice rooms, romantic. Shall I make a reservation?"

"I guess so," I said.

Mason could hear I was less than enthusiastic. "How about I book it for Sunday night after the kids leave? That way you can get more time with them and the warm water of the gulf."

"Thanks, Mason."

I got off the phone and made the kids hot dogs for lunch. Not the healthiest meal but an easy one and something they loved. The beach was meant to be an easy time for everyone. We ate sitting on the deck. The boys decided they wanted time to themselves, so I stayed on the deck and hunted for dolphins.

The boys returned an hour later in their bathing suits.

"Can we swim now?" Jason asked.

"Sure. Give me five minutes to change."

We headed for the beach and the water. Jason had become a good swimmer. Noah a little less so. The waves had picked up some from the recent storms, so we stayed close together. After we'd all had enough, we headed inside, showered and settled in the living room to play a game.

"I've had so much fun, Dr. Brown," Noah said. "My mom says she's sad she can't stay here, but she promises we'll come back."

"It's a long drive for her," I said. "Maybe she could stay over one night. I'll call her."

I left to make the call and returned to find the boys talking . . . of all things about the murder.

"Your mom will come early tomorrow and spend the night. Mason and I will stay in Apalachicola, so they'll be plenty of room. Now, what is it you've heard about a murder?"

"The boys were talking about it when we played soccer," Jason said. "They said some guy was pushed off a boat and drowned. Do you know about that, Mom?"

"I know a little," I said.

Noah and Jason turned to me. Before I could say more,

Lucie and Hannah ran into the house and found us on the deck. Mason came behind with cartons of food. "The girls wanted to eat dinner at home."

"Perfect timing," I said.

"Need to reheat it, back in a second," Mason said.

"Mom's telling us about the murder," Jason said.

I realized I was no longer Mommy—at least not when other people were around.

"Oh!" Hannah's eyes grew large.

"Uncle Mason wouldn't tell us a thing," Lucie said. "Please, Mama. You've always said it's better to know than to guess about things."

Now, I was in a pickle. I waited until Mason returned with the food. He placed the containers on the outdoor table. Lucie ran to get mats, and I went in to gather utensils.

Mason took orders for drinks and returned with a pitcher of iced tea and one of lemonade.

"The boys heard about the murder," I said, once we were settled with our drinks and food. "They want to know more. How did you girls hear about it?"

"Lurleen talks kind of loud sometimes," Lucie said, "when she's excited. Danny shushed her before we could hear much. Please, Mama, we're not little kids anymore. Hannah and I want to know what's going on."

"And so do we," Jason said.

"You want to handle this, Mason?" I asked, "now that you are officially part of this family."

Mason dipped a fried shrimp in cocktail sauce, ate it and started talking.

"Okay. Well, when I was raising my boys, they didn't ask so many questions about what we adults were up to, but I can see I have a more curious audience on my hands. Here's what's going on."

He stopped to eat another shrimp and take a sip of beer.

He passed around the plate of french fries. It was clear he was stalling for time. No one said a word including me.

"You're not going to help me out here, are you, Ditie?'

"It's your show, sweetheart, and I'm all ears."

"Yeah, okay. It seems a man named Brad Arborough fell off a boat a few days ago. They found his body on the shore—not far from where we had the wedding ceremony. We decided we didn't want to stay near such an unhappy event."

"You make it sound like it was an accident," Lucie said, "but Lurleen said it was murder."

"She's right," Mason said. "He was hit over the head and dropped in the ocean after he died."

I watched as Jason, Noah, Hannah and Lucie looked at Mason wide-eyed.

"Lurleen didn't tell us that," Hannah said. She shivered a little. "Did they—did they catch the murderer?"

"They think it might have been done by two men, strangers apparently, who asked for a charter on Brad Arborough's boat."

"A charter?" Noah asked.

"That's when someone wants to rent a boat to go fishing or take a trip somewhere," Mason said. "Like what we did when we took a boat up the estuary in Apalachicola."

Heads nodded.

"But why would those men kill this guy?" Lucie asked. "Did they want to steal his boat?"

"That's a good question," Mason said. "They didn't seem to want to steal his boat. No one saw them on his boat. Some kids found the boat adrift in the ocean, near the boat dock. No one was on board."

"So how do they know these men killed this Brad person?" Hannah asked. "Did the men confess?"

Now, we were getting into deep water so to speak. I waited to hear how Mason would handle this.

"I'm not sure the police know what happened. They haven't caught the men yet."

Hannah shivered. "So these bad men could still be around?"

My turn. I put one arm around Hannah. "We haven't heard if the men have been found yet or not, but you don't have to worry, Hannah. We are all here together and these men don't want to harm strangers. If they did kill Brad Arborough, they did it for some specific reason. None of you has to be frightened, and we will all be in this house together, doors locked."

"Maybe we should go inside right now," Hannah said. She was thirteen, a year older than Lucie, but, at this moment, she sounded much younger. Both girls had been through a lot of trauma in their young lives, but their ways of coping were different. Lucie always wanted to know what was going on, and she didn't seem to get scared about much. Hannah wanted to be reassured. She used to put on a tough girl act, but she didn't do that anymore.

"We can go inside," I said, "but you don't have to be afraid. No one is coming after us, and we—all the adults—will keep you safe. You have to worry more about stingrays and jellyfish than about the man or men who hurt Brad Arborough."

"It's true," Lucie said. "Those men don't want to hurt kids and Uncle Danny and Uncle Mason will keep us safe. So will Mama. "

I wasn't sure if I was an afterthought so my feelings wouldn't get hurt. I looked at Lucie. She smiled at me.

"Girl power," she said and pumped her fist.

The boys were quiet.

"Everyone okay?" I asked.

"Could we go inside?" Noah asked.

We went in and closed the curtains to protect the sea turtles. And maybe ourselves.

I searched through the downstairs collection of DVDs and found "Toy Story."

"Are you guys up for an oldie but goodie?" I asked. I showed them the cover.

"Sure," Lucie said, as if she were the mother who knew her brood needed to be comforted.

The kids settled on the couch. Mason and I sat on stools near the kitchen counter.

"I made a mess of that, didn't I?" Mason said.

"Not at all," I said. "If you hadn't answered their questions, they would have been more frightened. We'll stay close to them and they'll be fine. "

We all jumped when the front door opened an hour later and Danny and Lurleen walked in.

"That wasn't exactly the reaction we expected," Lurleen said. "We brought dessert. Is anything wrong?"

"Nothing that dessert won't fix," I said.

15

Lurleen and Danny had managed to bring home ice cream in a cooler Danny always kept in his car.

"It's a little melted but not bad," Lurleen said. "Chocolate, vanilla, coffee, and mint chocolate chip."

Lurleen knew my favorites about everything. In ice cream it was mint chocolate chip and coffee. I took orders from the kids.

"You two sit down and Mason and I will handle the hard work of dishing out ice cream."

Lurleen chose a large chair that sat in one corner. Within seconds, Hannah had settled in beside her. Danny sat between the boys. Lucie came to the kitchen to help.

"Everyone's scared, Mama," she whispered.

"I know. I'm glad Noah's mom is coming tomorrow. I think it's time you kids go back home, don't you? It's been a wonderful week, but I think we've had enough sun and beach and water."

"You're coming too, Mama?"

"Soon."

"I don't want to leave you here," Lucie said.

Lucie's instincts to protect the people she loved came out in full force when she sensed danger.

"It's okay, Lucie. Mason and I are still on our honeymoon, remember? I think we might spend a couple of nights in Apalachicola—we'll be safely away from whatever is going on here. That's why we moved out of the house so near where the body was found."

I'd learned long ago that being straight with Lucie was the only way to be.

"Will Lurleen and Danny move to Apalachicola too?"

"I can never say precisely what Lurleen will do."

We both laughed about that.

"Noah said his mom really wants to have us stay with her for a few days," Lucie said. "All four of us kids. She told Noah she was lonely."

"I think that will be great," I said. "I spoke to Hannah's grandfather and he offered to take everyone out to dinner when you got back to Atlanta. I hope this —" I searched for the right word—"this bad event doesn't make you forget what a wonderful wedding we had."

"It doesn't, Mama. I want to come back here every summer and celebrate."

"Me too, and we will."

I handed Lucie two bowls of ice cream. She gave them to Hannah and Noah.

"Nothing for me," Lurleen said. "I'm chock full of fish and chips. I couldn't eat another thing—unless it was one of your homemade cookies, Ditie."

I checked the freezer. "All my cookies are gone."

Lurleen looked a little guilty. "You can't expect to leave cookies in the freezer that I don't find and eat. And Danny is worse than I am!"

"I don't mind. It's supposed to rain tomorrow. If it does we'll make more cookies. I brought everything we'll need."

It stormed the next morning, and we made dozens of cookies. Noah's mom arrived around noon.

"Hi Melissa," I said. "You made good time. Not great weather I'm sorry to say."

"I got an early start," she said. She hugged Noah long and hard. "I've missed you. Did you have a good time?"

"The best." Noah said.

"Ditie, if you don't mind I think I'll take the kids back with me this afternoon. I looked at the weather for Cape Azul, and it's suppose to rain for the next few days."

"That makes sense. Are you fine about keeping the kids at your place for a while?"

"I'm looking forward to it. With three boys, I'm used to a crowd, and I love the idea of two girls in the house as well. I can keep them all week if you like. Then you and Mason can have a proper honeymoon."

I sent the kids off to pack. Lurleen would supervise the girls and make sure they left nothing behind. Mason would handle the boys. When everyone was upstairs, I offered Melissa some iced tea or coffee.

"Coffee," she said.

"The kids may be a little jumpy," I said. "We had to tell them something about the murder."

"I appreciate your giving me a heads up. It's one reason I came down here as fast as I could. I know Mason's a detective, but I don't like to have Noah exposed to any of this. I want my kids to grow up being able to take care of themselves. I know that's what my husband wants as well, but any threat of violence, that I don't want."

"Your husband is still away at sea?" I asked. He was in the Coast Guard and that meant some long deployments.

"Another month. I guess that's also why I came early and why I'm so happy to fill the house with kids. My older boys don't get back from camp for two more weeks. Frankly, the house is too quiet. You come back whenever you want.

Hannah's grandfather and great aunt will take over if I need a break."

"I'm glad to hear that, and I'm also glad to get the kids away from this. This appears to be a targeted attack against one man, but even so I'm relieved they'll be going home with you."

Danny joined us in the kitchen."Lurleen and I are in close touch with the Chief of Police Scott Huxley. He's convinced this is a one-off. Someone targeted Brad Arborough. We don't have a serial killer on Cape Azul even though we don't know why Arborough was murdered."

"That's a relief," Melissa said. "I knew you'd never put the kids at risk, Ditie, but I felt out of control with all of this. I couldn't protect Noah when I was so far away."

We hugged and that was the end of it. The kids seemed as eager to be gone as we were to have them leave. Everyone except Lucie.

"I don't want to go," Lucie said as she was hugging me goodbye. "I could help you the way I did before. I'm not afraid."

"I know that, Lucie. You are my brave girl, and you always have been. I won't be surprised if you become a detective like Mason. But for now, I want you to be my almost twelve-year-old daughter that I don't have to worry about. I'll keep you informed, and you can keep taking notes in that notebook of yours. We'll talk every day, and I'll listen to any thoughts you have."

"You promise?" Lucie asked.

"Of course. Now, you can help me pack a picnic lunch for your long drive back. Please call me as soon as you get to Noah's house and settle in. I don't care how late it is."

I sent the kids home with a picnic basket full of sandwiches, fruit, and fresh-baked cookies. By two o'clock, Lurleen, Danny, Mason and I had the place to ourselves. The storm had turned ugly and dark. The deck was swamped with water and we could see sheet lightning outside.

"I hope they can avoid most of this driving back," I said.

Mason checked the weather on his phone. "Atlanta's clear, and it looks as if this storm is local. They should be fine."

I let out an enormous sigh. "I didn't realize how relieved I'd feel about getting the kids off Cape Azul."

"Me too," Lurleen said. "We know a little more about the death of Brad, but none of it fits together well."

"I think I need a drink before you tell us what else you know," I said. "Something comforting."

"How about a Baileys and cream?" Lurleen asked. "Or a Kahlua and cream?"

Kahlua and cream always tasted like a milkshake to me. That's what I needed.

Lurleen fixed a drink for me and one for herself. The men declined the offer.

"I'll get a beer with dinner," Mason said. He sat beside me on the couch, his arm around me. I sipped my drink, and we all listened to the pounding rain.

After a few minutes, I said I was ready to hear Lurleen's new information.

"First off, Danny's agreed to help Scott with the case," Lurleen said.

"Why?" I asked.

"You know the line about keeping your friends close and your enemies closer," Lurleen said.

"You think he's the enemy?" Mason asked.

"Don't know," Danny said before Lurleen could respond, "but he seems pretty interested in Lisa and a little disinterested in solving the murder. That's one reason I offered to help."

"But if he had anything to do with the murder," Mason said, "he'd never want your help."

"Right," Lurleen said. "For now we're calling him a frenemy."

Danny nodded. "I like the guy, so I hope he's not directly involved, but I think he is involved with Lisa. I was surprised he said he wanted my help—maybe he didn't want his folks to dig deeply, and I'm an outsider. He may think I'll keep things quiet no matter what I find."

Lurleen smiled. "I kind of led him to believe that," she said. "I told him Danny liked to know what was going on, but he didn't feel a need to broadcast what he found to everyone. I kind of implied Danny didn't like the law enforcement system. I might have suggested he got on the wrong side of it one time."

"What?' Danny asked. He turned to Lurleen. "You implied I was a criminal of some sort?"

"Not a criminal, *chéri.* I might have said you left the police force under a cloud, something like that."

A burst of thunder, followed by lightning one second later ended Lurleen's statement.

"Retribution for your lies, Lurleen," I said.

Danny didn't look happy. Time to shift the topic. They'd sort things out—they always did.

"Will you come to Apalachicola with us?" I asked.

Danny shook his head. "Too far away."

"Then maybe we shouldn't go either," I said. "With the kids gone, there's plenty of room for us to stay here."

"I was afraid you were going to say that," Mason said. He looked as unhappy as Danny had.

"Safety in numbers," I said. "If we left now, we'd worry and be in constant contact with these two. Wouldn't it be better if we stayed in the same place with them? Besides, the sun is bound to come out at some point, and I want a little more time in the water."

Mason sighed. "Now, *I* need a drink."

16

Danny gave Mason a beer and got one for himself.

"All right," Mason said, "let's hear what you know."

Danny looked at Lurleen and she looked back at him, one eyebrow raised. "You had the last conversation with Scott," she said. "You start."

"As I mentioned, Scott doesn't seem too interested in this case. He said the guys who did it are long gone from Cape Azul by now. The bartender who saw Brad talking to two men told Scott the men were strangers, not from around here. He could tell that by the way they talked and by how they were dressed."

"Which was how?" I asked.

Danny shook his head. "Scott wasn't specific about that. I've got a follow-up with the bartender this afternoon. Wanna come?" This was directed to Mason, but Lurleen and I responded.

"Yes," we both said.

Mason took another sip of beer. "If everyone's going, I'll go along. I'll try to keep things in check."

"I don't think I need you as a peace keeper," Danny said,

"but I could use your skills as a detective, make sure I'm not missing anything. It's the place close to the entrance to the state park."

Mason nodded. "Ditie and I went there. Blue Cafe, bar in the front, dark."

"That's the place. According to Scott, that's where folks go if they want a private conversation."

"So, is this a place where drug deals are arranged, that kind of thing?" I asked.

Danny shrugged. "According to Scott, Cape Azul isn't the prime location for that. There are plenty of other spots along the coast where boats can arrive unnoticed and meetings set up on some unused wharf. This bar is more a place for wealthy folks, celebrities who want to have their private space. He described the Blue Cafe as a place to meet privately and not be seen."

"We saw Lisa and Wayne there," I said, "tucked away in a corner, looking very engaged with one another."

"It was Lisa who told me about what goes on there," Lurleen said, "that it was a place for paramours and mistresses to have a drink with their wealthy sugar daddies."

"Paramours?" I said. "Mistresses? Are you getting formal in your old age?"

"Old age?" For a moment Lurleen looked horrified. She stood, walked to a mirror on the wall and ran a hand over her face. "Oh, Ditie, you're teasing me. But you shouldn't joke about such things. Those words are accurate. According to my conversations with Lisa, a lot of celebs buy here so they can see who they want with no interruptions from the press or their public. You saw all the gated communities near the park."

"I wasn't sure those kept anyone out," I said. "You can come by foot along the beach easily enough. What celebrities are we talking about?"

She rattled off a few names—some I knew but most I

didn't. I was not a groupie. Lurleen wasn't either but she did know names and the gossip about them.

"Do these celebrities have a place in this story?" Mason asked.

"Scott says no," Lurleen said, "but Lisa says maybe. As we've all heard, Brad got around, especially to women with money. Lisa wondered about one woman in particular—Yuki Tillman."

"I don't know her," I said. "Is that her real name?"

Lurleen shrugged. "Can't answer that. She's an up and coming singer. Pretty good actually. She made a lot of money in Europe and she's now catching on in the US. Anyway, she's twenty-five, gorgeous, long blond hair. She owns a house near the park and seems to be special friends with the owner of the Blue Cafe. She even sang there once in a while. That's how she and Brad met."

"You got all this information from Lisa?" I asked.

'Most of it," Lurleen said. "Some of it from the owner of the Blue Cafe. He's another one of those people on the 'I hate Brad' list."

"That sounds like a long list," Mason said.

"It is, *chéri*, but there are other folks who claim Brad was misunderstood, that he really was a good guy."

"Where does Lisa Arborough fall on that continuum?" I asked.

"She's a funny one," Lurleen said, shaking her head. "She's the person who's been most used by Brad, but honestly, I'm not sure she ever stopped loving him. I think if he'd agreed to come back to her she'd have taken him back. She seems to be the one most interested in finding his murderer."

"'Seems to be' might be the operative phrase," I said. "I saw her with Wayne Baskin and neither one of them appeared to be mourning Brad's death."

"She's the one paying me," Danny said. "Scott said his

department didn't have the money to pay for an outside investigator."

I held up both hands. "I don't get her," I said. "Not yet anyway. Does the list separate out into men who hated Brad and women who loved him?"

"Mostly but not entirely," Danny said, "and—as you say, Ditie—we don't know who's telling us the truth about how they feel."

"Lisa is the most obvious suspect," Mason said, "the person who seems to gain the most from his death. There's no divorce settlement if the husband in the divorce is dead."

"Even so," I said, "why would Lisa have you investigate if she were the murderer or had hired someone to do the job?"

"That's the question," Mason said. "It sounds as if Scott is happy to let this murder go unsolved. If Lisa Arborough wasn't pushing to find out who murdered Brad, all this could rest undisturbed—some strangers off the Cape murdered Brad for whatever reason."

"Exactly," Danny said. "That's the draw for me—I want to know who actually killed Brad."

"Me too," Lurleen said. "It's a puzzle with so many missing pieces."

"Are you in, Mason?" Danny asked.

Mason looked at me. "Is this how we spend the rest of our honeymoon, Ditie?"

"Not all of it, honey, I promise."

Mason looked at me again. "Maybe we could have one night in Apalachicola at the inn I booked. Have a nice dinner, a quiet evening together."

"*Absolument*," Lurleen said, as if Mason had been talking to her. "You have your time with Ditie, and we'll see you tomorrow morning. Could it be on the early side?"

"We'll get here tomorrow when we get here," Mason said.

"Fair enough," Lurleen said. "Danny and I have plenty to

do until you get back. We'll handle the follow-up talk with the bartender and let you know if we have more info."

"Stay safe," I said.

"Always," Lurleen said.

Mason and I shared an overnight bag. We didn't need much. We drove to Apalachicola, first talking about the murder and later talking about our lives together. Everything went according to Mason's plan—and mine, to be honest. One lovely dinner, one night alone in an old inn. We did whatever we wanted to do, and in the morning I felt rested and more loved than I had in my entire life.

We woke up late, had brunch at the inn and got ready to leave.

"Bookstore," I said, after we packed up. The bookstore in Apalachicola was one of my favorites, run by a woman who knew everything there was to know about mysteries and local history.

She greeted us as soon as we entered her small store. "I'm glad to see you again," she said to Mason. "Your girls were delightful."

"Thanks," Mason said.

"Where are you staying?" she asked.

"Cape Azul," I said, and her face darkened.

"I suppose you know about the terrible accident that happened there—the man who was found on the beach."

"Brad Arborough," I said, and the bookstore owner nodded.

"Only we heard it was murder," I added.

Again, she nodded.

"Everyone's heard about it," she said. "Accidents happen here, especially boating accidents, but murders are something else. We all knew Brad one way or another."

"Did you like him?"

She made a face. "It was hard not to like Brad," she said. "Did you meet him?"

"Our only contact with him was an unpleasant one. He wanted us to kick out our guests staying in the big house he and his wife owned."

She nodded. "Yes, Brad could be difficult, especially if he'd been drinking. And then there was the nasty divorce."

"You heard about the divorce?" I asked.

"In a bookstore, you hear about everything, but I wasn't talking about the current divorce. I was talking about the divorce that happened years ago. I remember thinking at the time someone might get killed. No one did then."

"We heard a little about the first divorce," I said. "It must have been ages ago. I thought Lisa and Brad had been married a long time—ten years at least."

"That's true, but his divorce from Shannon, the first wife, was really nasty. I'd only just moved here. For a while I felt as if I'd landed in the wild west. I was afraid to leave the store. I've always lived upstairs above the bookstore—it's close to a few local bars. I heard all about the threats back and forth—some of those took place on the streets outside the bars. Howie, the younger brother, and Brad would have at it right below my bedroom window—they'd threaten to kill each other over the estate and who got what."

She stopped talking for a moment. "I apologize—I'm not usually such a gossip. I think this murder has unnerved me. Wherever Brad went, trouble seemed to follow."

"We've gotten caught up in the murder ourselves," I said. "We rented a house from Lisa Arborough and got a call from Brad that he'd rented it to someone else. I'm Ditie Brown by the way, and this is my husband Mason Garrett."

"I'm Dawn, and I've heard about you. Newly married—congratulations."

She saw my look.

"Sorry," she said. "There are very few secrets around here. We know all about you. You are a real-life amateur sleuth, and

I admire that. I get my excitement from reading murder mysteries, but you get yours from solving them."

"Actually, I'm a pediatrician."

"I know that, too."

"What else do you know," I asked, "about the murder I mean?"

"A little—maybe a lot. We hear all the gossip here in Apalachicola, but sorting out the truth from someone's wild ideas—that's harder to do. I guess that's where you two come in, Detective Garrett. We've heard about you as well."

"Do you have time to talk to us?" I asked.

The bookstore was empty.

She nodded. "But only if no one else is in the store."

17

A customer entered the bookstore before Dawn could say anything. Mason and I shopped while Dawn helped the middle-aged man find a history of Apalachicola for himself and a children's book for his grandson.

Mason found *The Everything Kids' Soccer Book* for Jason. "Do you think there might be something in here that Jason doesn't already know?"

"Probably. It's a fat book, and Jason will love that you bought it for him."

"Lucie is going to be harder," Mason said. "She reads everything."

"We'll look for something brand new," I said. Instead I found a book entitled *13 Art Mysteries Children Should Know.* "This isn't new, but it combines Lucie's two loves—art and mysteries."

We brought the books up to the counter as the older man was leaving. Two more customers arrived as we were buying the books.

"This isn't going to work," Dawn said. "I have an assistant coming in over the lunch hour. How about we talk then? In the meantime, I think you might want to take a look at one of our historic houses—Brad lived there as a child and then with Shannon. It's pretty much gone to ruin, but I heard someone bought it with plans to restore it. I'll meet you there in half an hour."

She wrote out the address for the house.

We took our purchases and headed for the door, but before we could get out, a man shoved his way in.

"Hey, Dawn—you coming to the memorial service for my brother?"

We stood to one side and listened to the conversation.

"Hi Howie," she said. "I didn't know you were back in town. Is the memorial service already scheduled?"

"Mine is," Howie said. "I don't know what Lisa has planned if anything. She's probably happy to see Brad dead. But the rest of us—his real friends and family—are meeting in a bar tonight—up on the river—The River Catch. Can you come?"

"I'll try to make it," Dawn said.

Howie stood there for a moment and then turned his attention to us. "What are you two staring at?"

There was plenty to stare at. Howie stood there with a surfboard in one hand—I'm not sure how he got it through the door of the small bookstore. He was dressed in shorts and a T-shirt. His dark blond hair was shoulder length, a shaggy cut as if he'd done it himself. He was older than I was—in his forties I'd guess—too old to be dressed the way he was. He looked as if he'd been frozen in time as a twenty-something surfer.

Dawn intervened. "These folks rented Lisa and your brother's house for a wedding ceremony, so they know what happened."

"We're sorry for your loss," I said. "I'm Mabel Brown and this is Mason Garrett."

Howie continued to stare at us. "I've heard about you and

that friend of yours, Danny Devalle. Lisa hired him to help with the investigation of my brother's death. We don't need help with that, and we sure as hell don't need strangers poking into our private affairs. You should go home, all of you."

Dawn walked around the counter and stood between us and Howie. "You don't have to get yourself worked up, Howie. These people don't mean you any harm. They might actually be able to help you find out what happened to your brother."

"We're not some backwoods rednecks here," Howie said. "We can figure this out on our own."

"We're not involved in the investigation," Mason said. "We're here for our honeymoon—we moved out of the house, so we could get away from what happened."

"Hmm," Howie said. "If you were staying in the big house, then you saw a bunch of pictures of me with Brad. We were close growing up."

"We weren't staying in the big house," Mason said, "and we didn't see any family pictures when we walked through it after our guests moved out."

"No pictures? I'm sure Lisa got rid of them."

"Why?" I asked.

"Don't give me that. You know the story, everyone does. I was friends with Brad's first wife, Shannon. People read that wrong. They made an innocent kiss seem like a torrid affair. Lisa is the person who really caused the break-up between Shannon and Brad—it wasn't my relationship with her."

"The first divorce," I whispered to Mason.

"Lisa put you up to this, didn't she? She wanted all the money for herself, and when she couldn't get it in the divorce settlement, she decided to murder my brother."

"We didn't come here to do anything but get married," Mason said.

"Oh yeah, sure. Shannon told me that wasn't true—she heard all about you. She wanted me to keep an eye on you."

"Brad's first wife?" Mason asked. "Why is she involved in any of this?"

"She's not involved," Howie said. "She never thought she got a fair shake in her own divorce. Brad tried to buy her off with some diamond and emerald bracelet, and Shannon wouldn't settle for that. When she heard about the divorce from Lisa, Shannon thought she should be here. She always believed Brad had money stashed away off shore, and she thought that money might come to light in the new divorce proceedings."

"What would we have to do with that?" Mason asked.

"I guess she thought Lisa might have hired you to look into things—finances, stuff like that."

"I'm an Atlanta detective. I investigate murders not finances," Mason said. "There hadn't been a murder when we arrived here."

That seemed to stop Howie in his tracks. "Yeah, well, all I know is that Shannon wanted me to keep an eye on you, and Brad didn't want you here at all."

"We found that out when we arrived," Mason said, "when he tried to rent the big house to someone else."

"I'm not sure why you don't believe us," I said. "I've been coming to Cape Azul for years. It was the perfect place for us to get married, and that's all we planned to do."

Howie took a step back and didn't speak for a few seconds. "I saw you with your kids and the other people. I heard about the wedding. But I also know how devious Lisa is. She let you and your friends stay in my brother's houses so you could snoop around."

"Lisa told us the houses belonged to her," I said. "She told us she was the one who had the money in the marriage. She also said she hoped your brother would consider a reconciliation."

"That statement—that last one—is the only true statement she made. She never had money. It was my brother who had the money, the Arborough money. You check that out before you leave town—you'll find out the truth. Lisa may have wanted my brother back but more than that she wanted his money."

The man was no longer shouting at us. Maybe he believed what we told him.

"You want to know who murdered my brother?" he asked after a few seconds. "Lisa did it. When she couldn't get her way she hired some thugs to kill him."

"I would think you'd want those men brought to justice," Mason said, "and Lisa, too, if she hired them as you say."

"We'll get justice all right," Howie said, "but we'll do it our way. We don't need any outsiders looking into things."

Howie walked out the door and then stuck his head back in—"See you at ten, Dawn. Things won't get started much before then."

We followed him outside.

A young woman walked up the street toward the bookshop. "Hi Howie," she said.

The man plastered a smile on his face. "Hi Jen."

She walked over to him and gave him a hug. "I'm so sorry to hear about your brother. What a terrible accident."

"Yeah," the man said. "Terrible. I'll catch up with you later."

He strode off without looking back.

"I'm Jennifer," she said. "I work in the bookstore. You're the newlyweds staying in Brad's house, aren't you? Terrible for you too, I'm sure—his body washing up near his house. Well, I'm late—I need to get inside."

We watched her enter the bookshop. Dawn appeared a moment later.

"I'm sorry about all that," she said. "Howard Arborough is a trip even when he's not tripping—a lot like his brother—all talk but not much action. He doesn't actually get up to much beyond smoking pot and surfing. Why don't you follow me to the old mansion and we'll talk there?"

She climbed in her car, and we followed her half a mile up the main road to a large plot of land overgrown with weeds. She parked on the street and we did the same.

The 'old mansion' looked as if it had had a hard life. It was a wooden structure that appeared to be falling apart. A sign outside a fence with an open gate said, "Sold".

"Someone bought this?" I asked. "They plan to restore it?" I thought about the mansion Lurleen had bought a year ago. Compared to this place it had been in great shape. It needed extensive repairs but it had good bones. I wasn't sure this mansion had any bones at all. "It seems as if it might be more of a tear down," I said.

Dawn nodded. "I'm surprised it finally sold—it's been on the market for years. I don't know who bought it or what the new owners plan to do with it. The land is worth a little, but I can't imagine the house is. Let's walk around to the back and find a place to sit and talk. I brought my lunch. Are you two hungry? I can share."

"We're good," I said. "We had a late breakfast."

The house didn't look any better from the back than it had from the front. I imagined fifty or a hundred years ago it might have been a lovely house. Now, I wondered what was holding it up. It seemed as if a good storm would flatten it.

Dawn must have seen me staring at the house and shaking my head.

"It's a wreck," she said, "kind of like the Arboroughs who owned it. They were wealthy fishermen back in the day. Started their own company. You'd see their shrimp boats all over the place. That was fifty years ago. Then everything went south."

"How do you know the history?" I asked.

"I first started coming here twenty years ago. There was something magical about Apalachicola that made me want to stay. I moved here permanently five years later after a divorce. I loved it then, and I still love it."

"Howie Arborough wanted us out of here," I said. "I guess I'm surprised you're willing to talk to us."

"I'm not one of them," she said. "I'll never be one of them. I'm still an outsider and always will be. I don't mind that. I get on with the tourists and with the natives. It's mostly live and let live around here, but there's something about Brad Arborough's death that jarred me. We have boating accidents all the time. People drink and drive cars or boats, it doesn't matter which. I thought Brad was drunk and I wasn't surprised to hear he died. Then I heard it wasn't an accident—it was murder. That shook me up."

"You know the Chief of Police, Scott Huxley?" Mason asked.

She nodded.

"He's investigating the case," Mason said. "I guess you heard from Howie that a friend of ours, a private detective Dan Devalle, is helping him."

She nodded again. "Your friend will find out only what Scott wants him to find out. Don't say you heard that from me—I'll deny it. I have to live with these folks."

"We won't repeat our source of information," I said, "but why do you believe that? Do you think Scott Huxley is a bad cop?"

"Not bad exactly. He wants to keep the peace around here, but he'll do what certain people ask him to do."

"Like Lisa Arborough?" I said.

"What do you know about Lisa?" Dawn asked.

"Very little. Only that she rented us the two houses for our wedding. What do you know?'

"I know a lot more than that."

18

We settled on an old bench in a shaded area of the property. Dawn pulled out a sandwich and an apple. "I'm sorry to eat in front of you, but if I don't eat now, I'll be starving later. Most of my customers come in the afternoon."

"Please," I said, and Mason nodded.

Between bites she told us what she knew. She turned out to be a keen observer.

"Have you been coming here long?" she asked me.

"Ever since I went to med school in Atlanta. Cape Azul was the perfect getaway for me. This is only Mason's third trip. I'm not sure what I would have done if he didn't like beaches and the gulf."

"I think it would have been the end of me," Mason said, "but the fact is I love being near the ocean. We're a good match in that way."

"You look to be a good match in many ways," Dawn said. "Solving crimes—is that something you do together?"

"No, no," I said. "That's Mason's profession. I'm a pediatrician at a refugee clinic."

"Is there still a call for that?" Dawn asked. "I knew for a while refugees were being kept out of the country."

"It's opened up again," I said. "Now, we have many Afghan refugees and some Ukrainians. It's work I love."

"But it seems you also solve crimes. As I said, nothing stays quiet here. I know you said you weren't involved with investigating Brad's murder, but someone told me you were."

"Would you by any chance know a woman named Lurleen?" I asked. "Tall, slender, auburn hair and hazel eyes. If you saw her in your store, I don't think you'd forget her."

"Yes, I remember her. She's the one. She said her boyfriend was off with four children and she had time to kill. Then she got onto the topic of Brad Arborough's murder. She asked a lot of questions and had a lot to say."

I smiled. "Lurleen always has a lot to say, and sometimes she stretches the truth. Mason and I came here to get married, have a honeymoon—that's all. Lurleen and Danny are our best friends. The kids belong to us—two of them anyway—Lucie and Jason. The last thing we meant to do was get involved with a murder."

"I see. That's not quite how your friend Lurleen presented it. Perhaps, you don't want to know what I have to say."

This time Mason smiled. "You don't know Ditie very well yet, but rest assured she wants to know what you have to say. We got married and had a wonderful celebration. We sent the kids home to keep them safe. My friend Danny is on the case. So that means we're involved whether we like it or not. Once Ditie hears about a mystery or a murder, she has to know what happened."

"Not if it involves the safety of the kids," I said. "I'm cautious. They can't lose me, and I can't lose them."

"All right," Dawn said. "I don't see how this will put you at risk, but I've never been too good about identifying a murderer in a mystery I'm reading until the author tells me who did it. Here's what I know. Brad Arborough grew up in this family home. As the oldest son, he was supposed to take over the family business, but fishing wasn't his thing. It turned out he was a good salesman. He could convince anyone to buy what he was selling. He was successful as a young man and people around here liked him."

She paused to take a bite of apple and drink some water.

"He went off to school, Vanderbilt I think, got a degree in business, and then returned to Apalachicola. I don't know if you noticed in your visits that sometimes Apalachicola seems to be thriving and other years it appears to be down on its luck."

"I have noticed that," I said, "and I've never understood it."

"Me neither. I mean I get by," Dawn said. "It's a river town—maybe that means its fortunes slip and slide like the river it's near. I don't know. Anyway, Brad came back from Vanderbilt with a girlfriend, Shannon, and a vision that the town was about to expand exponentially. He and Shannon had a grand wedding that I still hear about.

"Brad saw Apalachicola as the destination point in the South, full of art, tourists, wealthy folks looking for a second home. He bought up a lot of land using his dad's money and then everything went bust. I wasn't here when it happened. I came to live here a year or two later. That meant I got my bookstore for a song."

She glanced at her watch. "I have to get back, so Jen can leave. Let me speed this up. The Arborough fortune went belly up. I was here to witness that. It was mostly Brad's doing—his enthusiasm that outstripped reality, but it was also new fishing restrictions. And Shannon liked fine things—so the story goes. A perfect storm. His parents finally moved out of the house to the Villages in Florida."

"The big retirement community," Mason said.

"Yes. That was a dozen years ago. They never came back, so I don't know how they're doing. Maybe they'll come back for Brad's memorial service, but as far as I know the parents and Brad were estranged. Both for what he did to their fortune and for how his first wife treated them."

"Shannon," I said.

"Yes, Shannon. Shannon Arborough was a piece of work. She treated the senior Arboroughs as if they were doddering old fools, and eventually she treated Brad the same way. As I said, I was here when Shannon and Brad divorced and I was relieved that everyone survived that.

"In the end there didn't seem to be much money to split. Most of it was debt that Shannon left Brad to handle. Like a lot of Brad's relationships with women, this one had an ebb and flow to it. He brought Shannon home as the love of his life, then they hated one another for years, and then in the last few years it seems as if they might have become friends again. That's the rumor I heard. I think Shannon might have retained ownership of this crumbling house, and she might be the one who sold it. She couldn't have gotten much money for it as far as I can see, and I can't imagine who would buy it."

Dawn stood.

"I'm really sorry. I have to go, and I haven't even gotten to the best part—the run-in between Shannon and Lisa. That happened right after Brad's body was found."

"I know you can't help it," I said, "but you're dropping us in the middle of a page turner."

She smiled. "I have a friend who makes jewelry and loves to talk. The store is called Lost Horizons, and her name is Mary. Tell her I sent you and she'll fill you in on the rest of the story.

19

We stayed seated on the bench and watched Dawn drive away.

"Lurleen mentioned that store and Mary, didn't she?" I asked.

"Seems like it," Mason said.

"It's not like we have to be anywhere at any time," I said. "The kids are out of harm's way. Maybe we could talk to Mary in the jewelry store and hear the rest of the story."

"You can't leave this alone, can you?" Mason asked. "For you it's a lark but for me—well, for me it's work. I didn't plan to be doing work on my honeymoon."

I turned and wrapped my arms around him. "If you want me to stop, I'll stop. We won't go near Mary's store—although I know you said you wanted to buy me something as a honeymoon gift."

"That isn't fair," Mason said.

"No, it's not." I kissed him. "I mean it, Mason. I don't really have any skin in this game—I can let it go. We can just be together, spend another night here if you want or go back to be with Lurleen and Danny."

Mason smiled at me. "Lets have a late lunch and talk about it."

"I'll call Lurleen and let her know we may or may not be coming back to their place tonight."

Lurleen answered on the first ring. "Stay there, have fun. We're good. Danny and I are off to check with a couple of folks who knew Brad well—bartenders as you might imagine."

"There's some kind of memorial service here," I said, "that Brad's younger brother, Howie, organized for ten p.m. at a bar in Apalachicola."

"We heard about it," Lurleen said. "If you stay in town we could all go."

"I don't think we'd be welcome—I can't speak for you."

"If it's like any other bar around here, it will be dark and we can avoid being seen. And if you opt out, Danny and I will probably show up. We'll fill you in on what we learn."

"Take care," I said, "whatever you decide. Howie didn't seem happy to hear that Danny was working with Scott to find the killer."

"We always take care, you know that."

I hung up and turned to Mason. "We have permission to stay here another night if we want to. Apalachicola has always seemed like such an interesting town—let's explore it."

"With or without the jewelry store?" Mason asked.

"Your call."

We had lunch at the Osprey Cafe. I started with what turned out to be the best shrimp cocktail I'd ever had. Mason had a shrimp bisque, equally delicious. Then we shared a plate of fish and chips. Mason had a beer and I settled for water. A glass of wine in the afternoon heat would have made me ready for a nap, and I didn't want that.

As we ate, I could see Mason relax. I could also see his curious mind begin to turn. "There are so many things about this case that don't make sense," he said.

"Like . . . ?"

"Like how the Chief of Police Scott Huxley is handling the investigation. He appears to be a friend of Lisa Arborough's. She seems to want desperately to know what happened to Brad. For what reason we don't know. Scott stays lukewarm about the investigation."

"I agree," I said. "I'd say cooler than lukewarm—he's willing to place the blame on two strangers who may or may not have hired Brad and his boat on the day he died. He doesn't seem anxious to track them down."

"Right, Ditie, and yet he allows Danny to look into things —even if it's Lisa who's actually paying Danny. I can't understand why he's agree to that."

"Does he think Danny is incompetent?" I asked. "Is it a way to placate Lisa, get her off his back?"

"Anyone who's talked to Danny for five minutes," Mason said, "and checked his background would know how competent he is."

"You're right. It doesn't make sense. Of course, all our information is coming to us second hand. We haven't talked to the Chief directly about what he's doing."

"Except when he told us the break-in had nothing to do with Brad's murder," Mason said. "How could he know that?"

"Does this mean we're back on the case, Mason?"

"We were never on the case," Mason said, "but I want to understand things more."

"Does that mean we talk to the jewelry store owner?" I asked.

"Maybe. Let's wander through town and if we stumble onto her store, we'll go in."

I knew Mason was teasing me. He didn't leave anything to chance. "Okay," I said. "Let's explore the town and see what we find. But it's after three pm and I don't know when the stores close."

"You win. We'll find the store—Lost Horizons—and then we'll explore the town."

We asked the waitress if she knew of the store.

She laughed. "Everyone knows Lost Horizons—it's across the way and down the first side street you come to." She pointed us in the direction we'd want to go once we left the restaurant. "You can't miss it. Mary is a fixture around here. Stores come and go, but Mary's shop is always there. She makes the most beautiful jewelry. I have a piece." She fingered the necklace she wore—it looked like a collection of tiny seashells on a silver background. "I wear it all the time, especially when I'm away from this place. It reminds me of home."

"It's lovely," I said.

"She also promotes local artists. She has water colors, oils, if you're looking for something like that."

"We might be," I said. "Thanks a lot. Lunch was delicious."

"I'll tell the chef. He's new, and he'll be glad for the compliment."

We left a big tip and stood outside the restaurant for a moment to get our bearings. The town itself was little more than a main street with a few side streets heading off in both directions. It continued to have its historic charm, almost like a frontier town—only this one lay at the edge of a bay. Wooden false fronts stood tall along Main Street. The side streets contained lower buildings for the most part, and farther out of the small town stood Victorian mansions, many of them revitalized as bed and breakfasts.

"I do love this town," I said taking Mason's arm. "So much character, kind of like visiting with an old timer who has plenty of stories to tell."

We walked down the side street our waitress had directed us to, and we found Mary's shop. Tucked at the very end of the street was a sign that looked as if it had been weathered at sea—Lost Horizons.

I don't know what I expected—a wizened woman perhaps, someone much older than the person who greeted us.

"Hi," said a young woman who appeared to be in her early twenties. She was short—maybe my height but much thinner—with soft brown hair that curled around her face. "May I help you find something?"

"We're looking for Mary," I said.

"My grandmother. She's in the back working on her jewelry. Is there something I can help you find?"

"Dawn from the bookstore suggested Mary might be willing to talk to us for a few minutes," I said.

"About?" the young woman asked.

I hesitated.

"About some of the history of Apalachicola. Concerning the Arboroughs specifically."

"Oh," the young woman said. "You're here about Brad Arborough's death. You're not the first. It's a good thing my grandmother loves to talk or she'd be completely exhausted by now. Let me see if she can spare the time."

She left and was gone longer than I expected. Mason and I looked at the handmade jewelry as we waited. Every piece was unique.

"I'd love to buy you something, Ditie, but you don't wear a lot of jewelry."

"That's doesn't mean I don't love jewelry," I said. "Here's a brooch that's beautiful with a pod of dolphins frolicking in the sea. I'd like something that would remind me of our wedding and honeymoon."

"It's yours," he said without looking at the price.

I found the tag and frowned. "I want to support local business, but this is a lot of money for a souvenir gift that I will rarely wear."

"I don't mind at all," Mason said.

"Let's look at the water colors," I said. "Those we could see every day. Here are two I love. Same artist. They capture

Cape Azul." One showed a man walking on the beach with an umbrella on a stormy day. The other was a more abstract painting of sea birds, the shore, the mist. "I'll buy these. Then we'll never forget our time here."

The young woman came back before Mason could respond. By her side was a frail, older woman, slightly stooped but with bright, clear brown eyes. I could see the resemblance between the two despite the age difference of at least sixty years.

"Aren't these beautiful?" she said, looking at the two paintings I'd put on the counter. "Made by a dear friend. She'll be delighted you like them."

"They capture the coast here," I said.

"They do. Sadly, we are no longer the forgotten coast although we have managed to keep out big developers."

"I come every year to Cape Azul and visit Apalachicola. There are always more houses on the Cape but no high rises or big box stores around here. You've done a good job."

She nodded. "How can I help you? Chloe, my granddaughter, made it sound as if you wanted information about the death of Brad Arborough. I'm afraid I know nothing about that."

"Not so much about his death as about his past," I said. "Dawn, from the bookstore, suggested you'd know about the Arborough family and perhaps something about Brad's first wife, Shannon. She specifically mentioned an encounter between his first and second wives—Shannon and Lisa. It seems that happened after Brad's body was found."

"I know about all of that," she said, "but why do you have an interest in the family?"

Time to come clean. "Could we talk somewhere privately?" I asked.

"Back room. Follow me. Chloe, call if you need help."

The back room was a narrow space with a long wooden table that ran the length of the room. On it was silver jewelry in various stages of completion. Stools stood in front of the table.

"My work space," Mary said. "I have the stools so I can always sit when I need to. Arthritis. I have assistants now who help me. I'm hoping one of them will take over the work when I can't do it anymore."

"Chloe?" I asked.

"Chloe is a dear, but she has no interest in jewelry making or in living in Apalachicola. The town is too small for her, too small for most folks. Chloe doesn't mind coming here for her summer vacation—she can make good money in my shop. But to live here—no way. She'll be a Senior next year at the University of Alabama in Tuscaloosa, and I suspect she'll stay out there. That's where her mother has settled. Now, what is it you'd like to know and why?"

I explained to her about the wedding, the murder, our backgrounds.

She nodded along as I spoke.

"You are both detectives?"

"Mason is the real detective with the Atlanta police department. Danny is a private investigator—our friend. I've ended up involved in a few murders, but I'm actually a pediatrician."

"I called Dawn to check on you before I agreed to talk to you," Mary said. "That's why I didn't come out immediately. I don't keep up with current issues but Dawn does. She said she knew of you, Dr. Brown. She reads the Atlanta papers. You have a reputation for solving murders that sometimes elude the police. Commendable I suppose although I've never taken a particular interest in murders. They are always sordid affairs and sometimes the person killed deserves to be dead."

I hesitated. Was that a reference to Brad?

Mason stepped in. "I know your time is valuable. Could you start by telling us about the marriage between Shannon and Brad? It seems it was a stormy one that ended in a messy divorce."

"You got that from Dawn, didn't you? I tried to keep out of the way, but I got involved from start to finish anyway. No one wanted that marriage to happen. I was friends with the Arboroughs. They bought my jewelry. They thought Shannon was a gold digger, and I'm quite sure she was. Anyway, the parents finally agreed to the marriage. They wanted Brad to be happy, and he seemed to be mad about the girl. I made their wedding rings. Their wedding was a big affair. I don't think Dawn was here for that. Fortunately, I'd have to say. I didn't go. I don't care for those big social events, but everyone in town heard about what happened."

She stopped for a moment.

"I want to get the sequence right," she said. "I heard about it from Mrs. Arborough. It was twenty years ago, almost to the day. It seems everyone had too much to drink before the ceremony even began. According to Mrs. Arborough, Brad found Shannon kissing his younger brother, Howard, moments before the wedding, if you can imagine that. Not a friendly, welcome to the family kind of kiss—more like one that should take place in a bedroom, which I believe it did."

She paused to take a sip of water from a glass on the table.

"As Mrs. Arborough said later, if things had ended at that point everything would have turned out differently."

"The marriage, you mean?" Mason asked.

"Yes. According to Mrs. Arborough, Brad and Shannon should never had married."

Mary sighed.

"I hate conflict. Apparently, Brad beat his younger brother to a pulp. Then he went on with the ceremony. Mrs. Arborough said it was all Shannon's fault, but Brad didn't see it that way. It started a rift in the family that never healed. The brothers couldn't be in the same room together. And Shannon didn't make things easy for anyone. She was ruthless from what I heard. She took over the house, finally made the Arboroughs leave and ran their fishing business into the ground."

"Brad let her do that?" Mason asked.

"Brad was in love with her. He did everything he could to keep her happy, but in the end he couldn't do that."

"Dawn talked about a messy divorce," I said.

"Yes, just when it seemed things couldn't get worse, they did. Lisa came on the scene and now Brad was desperately in love with her. He initiated the divorce from Shannon. Every side brought in lawyers. Shannon, Brad, his parents, his younger brother, Howard. I'm quite sure most of the remaining estate went to pay off those lawyers. No one was happy in the end. Shannon got the house but not much else. Brad got nothing."

"The parents? The younger brother?" Mason asked.

"The parents had some claims. I think they managed to eke out enough money to live in a retirement community in Florida. I don't know what happened to Howard. He disappeared after the divorce—didn't turn up again until recently."

"You didn't know him well?" I asked.

"No, I knew his brother better. Brad liked my jewelry. I made a piece for him just before the first divorce. I didn't know who he planned to give it to."

"The piece?" I asked.

"It was unusual. I don't normally work with fine stones, but Brad gave me the emeralds and diamonds. I made him a bracelet. I don't know what he did with it."

"A diamond and emerald bracelet," I said. I could hear myself sounding breathless "Would you recognize it if you saw it again?"

"Of course," Mary said. "What artist wouldn't recognize her own work?"

20

"You look a little shocked," Mary said to me. "Why?"

"A bracelet like the one you described was found in a house near where Brad's body washed ashore—the house he and Lisa apparently owned together. Did you know that?"

"No. I'd be happy to take a look at it, although why it would turn up now after twenty years, I can't imagine."

I found the picture on my phone.

"This isn't my work," Mary said. "It's much grander than what I do—and I don't work in gold."

"Thank you," I said. "We understand Shannon has been in Apalachicola recently, and that you may have witnessed a confrontation between her and Lisa."

"Dawn told you that, but it was actually my granddaughter, Chloe, who saw it. She reported it to me. I don't get out much and I'm no longer interested in gossip. I spend the day in my studio if I'm feeling well and upstairs in my apartment if I'm not. Chloe can probably tell you about it word for word."

Mary stood and walked to the curtain that separated the

front of the store from the backroom where she worked. "Chloe, can you come back here for a moment?"

Chloe finished with a customer and joined us in the backroom. "What is it, Grandma?"

"You know all the gossip about the murder of Brad Arborough, and you saw what happened between Lisa and Shannon. These folks want to hear about it."

"There wasn't much to it, Grandma."

"Was there a scene?" Mary asked. "There always seems to be a scene when Shannon is around."

"They actually seemed to be commiserating over Brad's death—they hugged each other."

"That's odd," Mary said, "I didn't think the two of them could be in the same room together, and the idea of their hugging one another—that's almost unimaginable unless one of them had a knife with plans to stab the other." She chuckled. "Murder seems to make strange bed fellows. Thanks, Chloe—you can get back to the customers."

"We won't keep you a lot longer," I said, "but do you know much about Lisa Arborough?"

"Almost nothing except that it was always clear that Lisa and Shannon hated each other. I know Lisa comes to Cape Azul most summers and spends time in town. She seems to like my jewelry and may have bought a piece of two."

I nodded, thanked her, and Mason and I went back to purchase our water colors from Chloe.

I decided to show Chloe the picture of the bracelet. "Have you seen anything like this? Your grandmother told us about an old diamond and emerald bracelet she'd made, but she said this wasn't it."

Chloe looked at the photo. "This doesn't look like her work. And if she says it wasn't her work, then it wasn't." She handed me back my cell. It seemed to me she had more to say.

"What is it?"

"It's probably nothing. I heard from a friend that a bracelet was found in Brad's house on Cape Azul. Is this the bracelet?"

"Yes. What do you know about it—or what does your friend know?" Mason asked.

"She said she thought Brad was going to give her a bracelet like that. I said she must be out of her mind. I knew Brad didn't have any money, and they'd only been dating a few weeks."

"The friend's name?" Mason asked.

"I don't know that I can tell you that. She told me in confidence. She didn't want Lisa to find out they were dating."

"It's likely Scott Huxley already knows her name," I said.

"Oh my god, I hope not," Chloe said. "He'd have a fit."

"Why?" I asked.

"I've already said too much."

Time to let the matter drop.

"You don't need to buy me the brooch," I said to Mason. "I'm not sure when I'd wear it."

"I want to buy it for you," Masons said. "You can wear it when we go out to dinner."

Chloe wrapped up the two pictures and put the brooch in a box. I paid for the pictures and Mason paid for the brooch. We hadn't worked out our finances yet. We'd pool our resources, but for our small loves, we'd have our own money. I loved art. Thankfully, we had enough money to indulge our private interests.

Mason opened the box. "Wear it now?"

"Absolutely."

We thanked Chloe and called out a goodbye to Mary. She appeared at the door of her workroom, so I showed her the brooch.

"Funny story about that brooch. I always loved that particular piece. I made it a year ago—for Lisa Arborough as a matter of fact. Brad wanted to give her something special or so he told me. She brought it back a few days later. She said it symbolized empty promises, so she wouldn't wear it."

So many interwoven threads.

"Well, I love it. I won't be bringing it back to you."

We said goodbye and walked out the door. We stood under a porch that covered the entrance to the store. A rain cloud darkened the sky directly above us. I was always amazed at how rain could erupt in one place on the gulf and leave another spot one hundred yards away clear and sunny.

"Shall we wait for the downpour or try to make it back to our hotel?" Mason asked.

"Let's wait it out," I said. "I have a hundred questions I want to sort through."

We sat on the bench outside the store and watched people enter and leave Lost Horizons.

"Do you mind wearing a brooch that was once given to Lisa Arborough by a murdered man?" he asked.

"Maybe it should bother me, but it doesn't." I fingered the dancing dolphins. "Everywhere we turn, it seems we find out a little more. Most of it still doesn't fit together."

"What's your biggest question?" Mason asked.

Before I could answer that, we heard a clap of thunder, and the cloud burst over our head. We discovered the porch was leaking and darted inside Lost Horizons to wait it out.

Chloe was inside talking to someone on her cell. She looked agitated and ended the call as soon as she saw us.

"I'll bet she was calling her friend," I whispered. "I wonder what friend that would be—she made it sound as if Scott Huxley would be very upset that whoever it was had a relationship with Brad."

Mason nodded. "I know my wife always got upset when she thought the boys were hanging around the wrong kind of girls. Do we know if Scott Huxley has children—daughters perhaps?"

"If he does, they'd be young," I said, "and Brad was pushing fifty from what I've heard."

"All the more reason to be upset," Mason said.

The storm ended as abruptly as it started. We left the store and found some open space to escape the dripping porch overhang. I called Lurleen.

"Do you know if Scott Huxley has any daughters?" I asked.

"What an odd question," she said. "He does. One. I met her. She's a lovely young woman around twenty or twenty-one I'd guess. Long blond hair, fit, with a surfer's body. I think her name is Erin. Scott introduced us and then said how upset he was that she'd dropped out of school with no plans to return. Erin seemed pretty upset herself—I couldn't really tell why, whether it was because Scott was angry with her or something else. So, why the question?"

I explained what Chloe had said to us.

"*Ooh, la, la,*" Lurleen said. "You think Brad might have been dating the Chief's daughter! That might make Scott mad enough to kill the guy—and then to kill the investigation. He could send Danny off on a wild goose chase to make it look as if he were conducting a real investigation. *Mon dieu*! Scott Huxley could have bribed the bartender to make up the story about two strangers wanting to rent Brad's boat. I'm going to tell Danny about this, and we'll have another chat with that bartender."

"Be cautious about what you do next, Lurleen," I said.

"*Toujours, chérie.*"

She hung up on me. Lurleen's "always" didn't mean anything.

"Mason, I think I've unleashed a terrible can of worms," I said. He hadn't heard most of the conversation. "Scott Huxley does have a daughter—Erin. She's twenty or twenty one according to Lurleen with long blond hair."

"Long blond hair—like the strands Scott found in the bedroom in the big house," Mason said.

"Exactly what I thought," I said. "Lurleen wondered if Scott might have killed Brad and bribed the bartender to make up a story about two strangers in town wanting to charter Brad's boat. She said she'd tell Danny, and they'd follow up."

"That could be dangerous," Mason said.

"I know."

Mason called Danny.

"I'm here with Scott," he said. "What's up?"

"It can wait," Mason said. "Call me when you're free."

Mason and I sat in our car—the one place that was dry and private.

"Now what?" I asked.

"Let's sort through a few things about this murder," Mason said. "We've wondered if Scott Huxley might be dragging his feet about this investigation, paying lip service to it, using Danny to make it seem legit and then sending him to follow up on some strangers a bartender mentioned."

"Strangers who may never have existed," I said. "It would all fit if Erin was involved with Brad, and Scott found out. Maybe he got so angry he killed the guy."

"Yes," Mason said. "That means we have to go back to Cape Azul now. Head things off before Danny—or more likely Lurleen—gets in real trouble."

"Why don't we get them to come here?" I said. "They could spend the night, and we could go the memorial service in that bar."

"That could lead to a confrontation with Howie," Mason said, "not the best idea."

"Better Howie than Scott," I said, "if Scott is the murderer."

"Agreed." Mason called Danny with the plan.

"We can do that," he said, "after I speak to Lurleen. Clear

our heads a little and see who turns up at the impromptu memorial service. You don't need to book a room for us. I won't be drinking and I can drive us home."

"It's The River Catch," Mason said. "We'll meet you there early so we can see who comes and goes. Maybe we can avoid being seen ourselves."

Mason and I headed back to our hotel for a short nap and a light dinner. We headed out to the bar a little after nine.

21

The River Catch was crowded when we arrived. We spoke to the bartender about a memorial service.

"Oh yeah," he said. "Howie's planning a big send off for his brother. Crazy really. They never got along as adults but I guess death heals all wounds. It will be in the back room, but he's not here yet."

"We'll sit out front," Mason said. "Wait until other folks show up."

We found a dark corner in what looked like a very popular bar. It seemed to be a fisherman's hangout with lots of folks talking about their catches for the day or grumbling about state restrictions. Mason and I nursed drinks until Danny and Lurleen arrived. We waved them over to our table moments before Howie showed up.

He headed directly for the back room. On his arm was a woman who looked remarkably like Lisa, except that she was older and more buxom. This had to be Shannon, I thought. Apparently, Brad had liked his women blond. That would make Erin Huxley a real possibility for his latest conquest. Scott Huxley found long blond hairs in the master bedroom of the big house—did he recognize they might belong to his daughter?

Howie had cleaned himself up. He'd tied his dirty-blond hair back in a ponytail, wore dockers and Birkenstocks with a clean white shirt tucked into his khaki pants. Behind him came an entourage of people, men and women, who looked to be in their forties and fifties. They were a mixed crew. Most were casually dressed and some looked worn around the edges—as if they spent a lot of their time in the sun, on the water or in the local bars.

A couple of men wore suits, which seemed out of place. They apparently thought so too. They shrugged off their jackets and loosened their ties. Howie greeted them as if he were particularly glad to see them. He shook their hands enthusiastically, but we were too far away to hear the conversation. He pointed them towards the back room.

"Everyone won't fit in the back there," the bartender said to Howie, "if this stream keeps up."

"That's okay," Howie said. "Let people sit where they want. The drinks are on the house. You got the mic set up?"

"Sure."

"Good, then everyone who wants to speak, can, and we'll all be able to hear what they're saying. I want this to be a special day for my brother."

Howie's game plan worked perfectly for us. We could stay in a dark corner of the bar and hear everything that went on.

For the first hour nothing much happened.

Dawn came a little after ten, alone. I started to wave her over, but Mason held my hand down.

"Better no one knows we're here," he said, "if we can manage to pull that off."

Lisa and Wayne turned up as the speeches were starting. They remained standing at the door to the back room, their backs to us. Moments later, Mary and Chloe came with Lenny and a tall blond girl beside him. Our mystery lover I wondered—Erin, perhaps, daughter of the chief of police. I expected Scott to be there, but he never showed up.

His deputy came, and he was the only one who spotted us. He came over to our table.

"What are you four doing here?" He asked.

"We were in Apalachicola and heard about this . . . memorial service," Mason said. "We thought we'd pay our respects."

"You don't really expect me to believe that, do you?" the deputy chief asked. "You didn't even know the guy from what you told us. I'm here to keep the peace, but why are you here?"

"To be honest," Lurleen said, "we're curious about Brad Arborough and want to know what people have to say about him."

"My guess is you'll be just as confused at the end of this as you are right now," the deputy chief said. "People loved him or hated him."

"You were friends with him?" Danny asked.

"I was smart enough not to be friends with him. I'm careful with my money. I didn't let him bamboozle me out of a fortune, not like some folks I could mention."

"Like Scott Huxley?" I asked.

"I don't get into Scott's business," he said, "and I'd recommend you don't either." He looked toward the back room. "I gotta go. If you're smart you'll get out of here before a brawl starts. I'm one guy—I won't be able to protect you from twenty angry men inside that room once they get good and drunk."

"Thanks, officer," Mason said.

"I checked you out, Detective Garrett. I know you came here for your wedding. You'd be smart to stick to enjoying your honeymoon—lots of folks don't like meddling outsiders."

With that, he walked away.

"Was that a threat?" I asked.

"Don't know," Mason said.

"I like the guy," Danny said. "I trust him more than anyone else in that police department."

"Meaning Scott and Lenny?" I asked.

"Yeah," Danny said, "meaning Scott and Lenny."

We got quiet as people started to speak. At first the comments were brief and polite. People, mostly men, spoke about how long they'd known Brad, about what he was like as a child, about the Arborough family. A few told funny stories about how Brad as a kid was a lot like Tom Sawyer—he could con you out of your last nickel without leaving you furious, often without you even realizing what he'd done.

Slowly, the comments got less polite.

"Why don't we cut the bullshit?" one man said. "Half the people in this room lost their life savings to Brad. We got nothing to show for it. That's why we're here isn't it? To see what little brother Howie is going to do to fix things."

I didn't know who the speaker was, but I did recognize Howie's voice.

"I am not my brother's keeper," he announced. "I thought Brad deserved a send-off even though he bilked me out of as much money as he did the rest of you. He's dead, man, there's nothing any of us can do about that."

Someone grabbed the mic. "You could give us our damn money back—that's what you could do, Howie."

"I don't have the money," Howie said. "You think I'd be living like this if I had the money?"

"We heard you bought back the family estate," someone yelled. "So you must have some money."

Howie responded. We heard everything he said loud and clear. "You've seen that house. Someone practically paid me to take off their hands. It's got sentimental value for me, and I don't even have the money to fix it up. I grew up there. I want to live in that house. If I had money to give you for my brother's mistakes, believe me I would."

People in the room started to grumble—it felt like the beginning of an earthquake.

The two men in suits left the back room. It seemed they couldn't get out of that bar fast enough.

"Do you know those two?" I asked Danny.

"Nope. Not Howie's's type I'd say."

Things turned ugly fast. Before they did, I saw Mary and Chloe leave. Lisa and Wayne weren't far behind. So, at least they were out of harm's way.

People stopped waiting to get their hands on the mic. We could hear shouting, then tables being overturned. We caught a glimpse of the deputy chief as he came out of the room briefly to call for back-up. After that we heard him on the mic.

"You all settle down or you'll get arrested," he said.

"Time to get out of here," Mason said leaving money on the table.

The four of us left. Outside, we ran into Mary and Chloe.

"What are you doing here?" I asked.

"I have the same question for you," Mary said.

"Later," Mason said, as we began to see people stumbling out of the bar. "Meet us at the hotel," he said to Danny and Lurleen.

They did. We sat in our hotel room and took stock of what had happened.

"All we got out of that," I said, "was a longer list of people who wanted to see Brad dead."

"Maybe," Mason said, "but we mostly saw people who wanted their money back. Would you kill the guy who'd stolen your money or would you want to force him to give it back to you?"

"That could mean a threat turned into something more deadly," Danny said. "Do you think the murderer would have come to hear what people were thinking? To find out how close he was to being exposed?"

"I would think that," Mason said.

"But Scott wasn't there," Lurleen said. "He's our number one suspect, and he could have been there to keep the peace. You're not letting him off the hook are you?"

"No," Danny said. "His deputy will tell him everything that happened, so he didn't need to be there. But it is curious he wasn't there—he's the chief of police and he'd know a fight was likely to break out. I don't know what it means that he didn't come."

"What about the Suits?" Lurleen asked.

"Suits?" I asked.

"You know those guys wearing those expensive suits—the bureaucrats, CEOs, whatever they were," Lurleen said. "The ones who flew out of the room before real trouble started. Did they want their money back, too?"

"Good question," Danny said. "Maybe Brad burned the wrong people—people who could hire a hit man. Maybe these guys wanted to make sure the work was done without any incriminating evidence left behind."

"And yet Howie appeared to be delighted to see them," Danny said.

"Maybe he's the one who hired them," I said.

"I don't know many hit men who return to the scene of the crime," Mason said, "unless they had to clean up something that went wrong."

"I wish we'd been in the room," Lurleen said, "so we could have taken a good look at everyone."

"It's lucky we weren't," Mason said. "We'd never have gotten out of there one piece."

He stood. "This is enough for one night."

Danny took Mason's cue. "Come on, Lurleen, I'll drive you home. We'll leave Mason and Ditie in peace. Let us know when you're coming back, if you're coming back at all. It may be you want to stay here away from the fray."

"We may, but there's one more thing you should probably

follow up on," Mason said, "Erin Huxley. See if she was Brad's latest conquest and if Scott knew about it."

"I'll follow up," Danny said. "Discretely."

MASON and I took a long time to fall asleep that night, but once we did we were out to the world until late in the morning. I suppose it really was good to be away from the fray.

We had breakfast in our room and talked about what we might do next.

"It's peaceful here," I said, "and I still haven't seen as much of Apalachicola as I'd like. No more bars for me, however."

"I second that," Mason said, "so maybe we finish out our honeymoon here?"

"I think maybe we do. I'd like it better if we could convince Lurleen and Danny to come join us and leave the investigation to the police."

"Can't control other people, Ditie," Mason said, "especially if those other people are Lurleen and Danny."

"I know. Maybe we'll just tell them what a great time we're having here."

"Want to venture out or stay in for the rest of the morning?" Mason asked.

"In, for now," I said.

We stayed in our pajamas, read the local paper, and talked about our future together. Neither one of us said another word about the murder.

When we felt like it, we dressed and headed for the Gorrie Museum that Lucie had mentioned. I came away with some understanding of the ice machine he invented and patented in 1851. As a physician, I was impressed with the fact that what inspired him was a deadly yellow fever epidemic in 1841. He came to believe that cooling patients might be a way to break their fever and help them survive. Instead he created a device that let all us of us refrigerate our food and ultimately live comfortably in a hot environment. I was happy to give this man his due.

In his honor Mason and I settled on ice cream cones for lunch.

We returned to our hotel with plans to read and enjoy more quiet time together until we got a frantic call from Danny around four in the afternoon.

"I'm on my way home, and I can't reach Lurleen. She isn't with you, is she?"

"No," Mason said. "She's not with us. When did you last talk to her?"

"Late this morning. I told her I always needed to know where she was going, and she promised to do that."

"She didn't leave you a message?"

"Not on my cell. At first, I assumed she was out for a walk or in the water, but now I'm starting to get worried. I'm headed to the house."

"How far away are you?" Mason asked.

"Five minutes."

"Call us back when you get there."

"Will do." Danny hung up.

"Let's give Danny time to get home before we panic," Mason said to me. "Her phone might be dead or out of commission—you know how spotty reception is on Cape Azul. We'll give him five minutes to find out."

"Or she might have decided to do more of her own investigating," I said.

Danny called us back. "Not here," he said. She left her phone on the counter. Maybe she went for a swim, but I'm worried."

I grabbed the phone from Mason. "We're worried too. Don't move until we get there."

Danny must have seen us driving down the sandy lane to

the condo half an hour later. He ran down the porch steps to meet us.

"Hi," he said. "I may have brought you home unnecessarily. I found a note from Lurleen. She said she couldn't get phone reception and she had to go off with Lisa for the afternoon. I was not to worry."

I studied the note. "I wish she'd said a little more."

"Me, too," Danny said. "I don't have Lisa's cell. Do you? I assumed she had a business office on Cape Azul, but I couldn't find a number for that either."

"I have her number," I said. I called and it went to voice mail. I left her a message that I needed a call back. Then I texted her—'Is Lurleen with you? I need to speak to her.' I left my name in case she didn't recognize my number.

Lisa texted back. 'in the middle of something. I'll call when I can. Lurleen isn't here.'

"Now what?" Mason asked.

Before any of us could answer, Lurleen strolled in as if she'd been out for a long lunch with friends.

"What's all this, *chéris*? I thought you and Mason were staying in Apalachicola again tonight." Then she looked at our faces. "What's wrong? Has something terrible happened?"

"We didn't know where you were, Lurleen," Danny said.

"You didn't see my note? The phone service is terrible, so I left a note."

"I didn't see your note until an hour ago," Danny said. "It's after five. I didn't have a number for Lisa."

"I did," I said. "She said she didn't know where you were."

"Oh," Lurleen said, and that was all she said.

"Where were you?" Danny, Mason and I said together like a Greek chorus.

22

Lurleen must have seen how upset we were. "I'm sorry," she said. "I needed to talk to Lisa—about some things. She said we could have lunch, and then she texted back that she couldn't make it. I tried to call her to reschedule, but my phone had no service. I couldn't sit around doing nothing, and I didn't want you to worry about me."

"So you lied to us instead?" I said. "I was worried sick about you after last evening in the bar. You said how much you wished you'd been in the back room, and I imagined you decided to do more investigating on your own."

"Was that what you were up to, Lurleen?" Danny asked.

Lurleen seemed suddenly at a loss for words.

"So, Lurleen," I asked, "why did you want to talk to Lisa and where did you go when she wasn't available?"

"I did try to call you, Danny," Lurleen said, "but the receptionist told me you were with Scott. Since he's a prime suspect I couldn't leave a message. And since you had the car, I couldn't talk to the bartender at the Blue Cafe to see if maybe he made up the story about two strangers wanting to rent Brad's boat."

"Thank God for that," Danny said.

"So?" I asked.

Lurleen sighed. "So—I didn't know what to do with myself. I decided I'd take a walk up the beach to see the wedding houses—Brad and Lisa's houses—just for something to do. It's only a half a mile, so that's what I did."

"You walked there in the early afternoon and got back after five?' Danny said. "Sorry, Lurleen, that doesn't fly."

"It was after two when I set out," Lurleen said,"after I couldn't reach Lisa."

"That still doesn't fly," I said. "Look, Lurleen, you need to tell us where you went and what you've been up to this afternoon."

"May I have a glass of water and then a glass of wine?" Lurleen asked. "It's hot out there."

"Yes to water," I said. "No wine until we know what you've been doing all afternoon."

Mason brought her a large glass of water. We sat on the couch and waited as she drank it.

"Much better," she said when she'd finished. "All right. I did go to the two houses we stayed in. I thought maybe I'd run into Wayne and Lisa there. They seemed to be hanging out a lot together. I imagined when Lisa said she couldn't do lunch it might be because she was with Wayne.

"I thought if I found them together, maybe they'd talk to me—tell me what people looked like in the back room and maybe more about Scott—what kind of man he was, would he be capable of murder, did he know about the relationship between Brad and his daughter."

"Oh, Lurleen," I said. "When I talked to you, I begged you to be cautious and you said you would be. Then you go out to investigate on your own."

"You and Mason didn't think Lisa or Wayne was the murderer. You thought it might be Scott. I was simply trying to get more information on that."

"Lurleen, we don't know who the murderer is," Mason

said. "We got information that might or might not lead to the murderer."

"You said you'd talk to Danny," I said.

"I told you I tried to reach him. I couldn't very well tell him what you thought with Scott Huxley in the background."

Danny had been silent this whole time. "This could have been very dangerous for you," he said. "We don't know these people, and we don't know what kind of relationship they have with each other."

"I was on the beach with lots of people all around me," Lurleen said. "I didn't intend to go inside the house when I saw Wayne on the veranda."

"You didn't intend to go inside?" Danny said. "That means you did go inside."

I'd never heard him sound so angry before.

"It was broad daylight, Danny. Wayne waved at me. I waved back, and then he invited me up. I couldn't very well refuse to go."

"You most certainly could have refused to go," Danny said. "We don't know who murdered Brad Arborough or why. Sometimes, Lurleen, ..."

He didn't finish his sentence. He didn't have to.

"I'm sorry, Danny. I'm so sorry I worried you. It's just that I know Wayne isn't a murderer. He's a nice guy and he invited me inside and I went."

"Lurleen," Danny said, "you think you are such a good judge of character, and sometimes you are. But not always! You seem to think you can never be harmed—that you're invincible. You're not invincible."

"I know. I know. But when an opportunity presents itself, how can I say no? An opportunity that might give us all kinds of information."

Lurleen stared at Danny, and he stared back.

"I know what you're going to say, Danny. Curiosity killed the cat."

"No, I wasn't going to say that. I was going to say I can't continue this investigation with you on Cape Azul. You're a wild card, and I can't be worrying about your safety. Either you go off the Cape or I quit and we all go home. That's the deal."

I always wondered what might make Danny snap. This seemed to be it. Lurleen looked as if he'd slapped her. She looked furious and then she looked crushed.

"You can't order me around. I'm an adult. No one orders me around."

Mason and I looked at each other. "We'll be out on the deck," Mason said.

"You don't need to go," Danny said. "This has been a long time coming. If I'm out of line, you can tell me, Mason. I'll listen to you, but I'm done listening to Lurleen's excuses about why she did something so dangerous."

"Let's all have a glass of wine," I said. "Let's slow everything down. Lurleen wasn't hurt, Danny. I'm not justifying her actions, but no one was hurt. Everyone is okay. Let's talk this through."

I got a ping and a text at that moment. It was from Noah's mom, Melissa, to say the kids were home with her. I went out on the deck to try to get reception enough to call back. Mason was in the kitchen pouring wine.

I couldn't get more than two bars on my cell, so I texted back. 'so glad. no reception here. tell the kids I love them. I'll call when I can.'

I came back into the main room as Mason was handing out wine. "The kids are home, safe. I'm so glad they are out of this."

Lurleen looked at me crestfallen.

"I don't mean the argument, Lurleen. I mean the murder investigation."

Lurleen nodded and sipped her wine. I grabbed some peanuts and pretzels from the counter above the sink and put them on the coffee table. "Maybe we're all a little hungry. Shall I fix dinner?"

"Later," Danny said. "We need to hash this out."

"Are you sure you don't want some privacy while you do that?" I asked.

Both Lurleen and Danny shook their heads no.

I sat in a chair near the sliding glass windows that looked out at the sea. I let the ocean calm me as I sipped my wine and ate pretzels. I didn't even like pretzels, but it gave me something to do. My dearest friend was having a huge argument with the man she loved, and this was a problem I couldn't solve. Mason looked as unhappy as I felt.

As Danny said, this had been a long time coming. Sometimes Lurleen listened to Danny, but sometimes she didn't. Sometimes, Danny could come on a little strong when he got worried about Lurleen's safety. They'd have to work this out or else.

Or else what, I thought? Or else, this relationship wouldn't last. Danny was more than five years younger than Lurleen, but that didn't seem to be a problem for either of them. Danny didn't care about an age difference, and he didn't seem to care that Lurleen was gorgeous. He'd seen her at times when she wasn't.

For Danny, and I think for Lurleen—the two of them just fit together. I remembered a card I'd seen once. Two people with jagged edges were side by side like two pieces of a puzzle. 'Made for each other,' that's what the card said. That's what I always thought about Danny and Lurleen. They were made for each other.

I sighed. Maybe I was wrong.

No one spoke for a few seconds. Then Danny and Lurleen both spoke at once. Lurleen to say she was sorry. Danny to say he sounded angrier than he meant to.

Mason and I sat quietly in our separate chairs.

"You said this was a long time coming, Danny." Lurleen

looked close to tears. "Have you been storing up resentments about me?"

"Not resentments, Lurleen," Danny said. "Worry. Lots of worry. When you were converting the Atlanta mansion into a cooking school, I worried about how much danger you were in—someone was threatening you and we didn't know who it was."

"When you told me to go nowhere without you," Lurleen said, "I did what you asked."

"That time you did," Danny said, "and this time you didn't. How can I keep you safe if I don't know when you're going to listen to me and when you're not?"

"You can't keep another person safe," Lurleen said. "If that were possible I would have saved my mother from the bad men who used her and from the terrible fire that ended her life."

I wanted to go to Lurleen and put my arms around her, but it wasn't my place. Danny did it instead. He moved next to her and hugged her.

"I'm sorry, Lurleen. I've never loved anyone as much as I love you. I can't bear the thought of losing you. But I've never been with anyone as headstrong as you."

"Have you looked in the mirror, Danny?" Lurleen asked.

Mason and I shared a smile, both because we knew what Lurleen said was true, and because we shared that same trait of stubbornness.

Even Danny had to smile. "I've heard that before, Lurleen." He reached for some peanuts. "But that doesn't solve our problem. You act as if you can do whatever you want whenever you want. If we're a couple, we need to make decisions together."

"I've always made decisions on my own, Danny, you know that. You said you liked my independence, and now you don't." Lurleen turned to me. "You know me better than anyone, Ditie. You've never complained about my independence."

"You're my best friend, Lurleen, but you're not the person I live with. Mason and I have had to sort out our roles together. We're still sorting them out. It's a partnership, Lurleen, not two ships sailing in parallel across a stormy sea."

"Then you think I'm wrong."

"Not wrong," I said. "I think you and Danny sometimes see things in black and white. Loving someone isn't black and white."

I remembered another card I'd seen that said, "I love you so much. Well, maybe not so much today. Today you're annoying me."

"Mason and I will be working on our relationship for the rest of our lives."

"I guess that's the question," Mason said. "Do you two want to be together in spite of your peccadillos?"

"Peccadillos," Lurleen said. "You make our differences sound so small—little annoying traits. The fact is, Danny, I love that you want to protect me, that you want to keep me safe. I just want you to realize you can't do that. I can work on listening before I act, but you must learn to respect me as a separate human being with a mind of her own."

Danny nodded. "I know, Lurleen. I love that I never know what you'll say or do next. You make me laugh. And you make me worry. I'm a person who likes ground rules—one of my peccadillos I suppose. Can we work on finding some ground rules we can both live with?'

"We can work on that," Lurleen said.

"I stood up. "Maybe we could order pizza tonight and Mason and I will pick it up."

"That sounds wonderful," Lurleen said, "but first, don't you want to know what I found out from Wayne?"

23

I sat down again to hear what Lurleen had to tell us. We'd order pizza after she finished. Mason joined me on the love seat.

"We have some questions for you, Danny, as well," Mason said.

"You go first, Lurleen," Danny said.

"*Eh bien*," Lurleen said. "Ditie, you told me that Wayne seemed like a straight-forward guy when you first met him. You said he got upset when it looked like he had no place to live but he settled down when you explained the situation."

"That is what I thought, Lurleen."

"So, when I saw him outside on the veranda of the big house, waving to me, I thought he wasn't a dangerous criminal. I thought I could get some useful information from him, see what opinion he had about Brad and who he thought might have killed him. I planned to stay outside, but it was hot. He could see I was perspiring. He invited me inside and I went. Danny, you know all the work I've done with my personal trainer, Wendy, —I could take on anyone, even you I suspect."

"We're not testing that out, Lurleen," Danny said.

"No, of course not. My point is that if I had made a mistake about Wayne, I could get out of there in one piece."

"You couldn't run from a gun, Lurleen," Danny said.

"Wayne didn't look like the kind of guy who carries a gun," Lurleen said.

"I carry a gun," Danny said. "Mason carries a gun."

"You guys have an edge to you—Wayne looks more like a softie—someone who wouldn't have anything to do with guns."

Danny stared at Lurleen. "There you go again making assumptions about people that could be wrong. Fatally wrong."

"Point taken, Danny, but Wayne wasn't carrying a gun. He was in his swimming trunks for goodness sakes. I think he wanted to talk things through as much as I wanted to hear what he had to say. He spoke to me nonstop for two hours. That's why I didn't get back here until just now."

"Okay," I said, "I'm ready for the story. From the beginning, Lurleen."

I didn't need to tell her that. Lurleen always told her stories from the beginning and always with as many flourishes as she could insert until one of us begged her to get on with it.

"As I said, Wayne was on the porch as I walked by. He was in his bathing suit as if he'd just been out for a swim or planning to take one. He's not a bad looking guy—for his age and all."

"His age?" Mason asked. "He's probably my age, I'd guess."

"Right," Lurleen said. "No offense meant, Mason. I'm just saying when I first saw him he seemed kind of dumpy looking. That's not quite right. He looked like someone who didn't take care of himself. But with his clothes off, he looks pretty fit. And he's probably close to fifty."

She looked at Danny. "Of course he can't hold a candle to you, Danny. You're half his age and have twice the body he could ever have. All I'm saying is he looked fine, like maybe his time at the beach had been helpful to him."

"Or maybe it was being near Lisa," I said.

"Right," Lurleen laughed. "You can pick up on those romantic undercurrents, Ditie, and so can I. Wayne definitely has a crush on Lisa. He finally told me about it. Now, I'm getting hungry. Can we order the pizza, and then I'll tell you everything else."

Mason handled the order. We'd get it delivered. He sat next to me and nodded at Lurleen to go on.

"Right, so as I said Wayne invited me inside for a drink. I settled for water. I wanted all my wits about me, and I asked how he was doing. He told me he was doing great. Then I guess he thought better of it and tried to tell me how sad he felt about Brad, how they'd been such good friends.

"I reminded him that we already knew Brad owed him a lot of money and that he'd come down to collect what was owed him. That seemed to sober him up a little. 'That's all straightened out,' he said. 'Lisa says she'll give me the smaller house she and Brad owned together.' "

"Wouldn't that more than clear the debt?" I asked. "Even the smaller house has to be worth a lot of money. How much did Brad owe him?"

"Wayne talked about a loan of ten thousand," Lurleen said, "but there was more money he invested in Brad's schemes. In any case, you're right, Ditie. I had the same thought. Even if Brad owed him a lot more, he'd still get a good deal—if Lisa followed through."

"Assuming he wants a house on Cape Azul," Mason said.

"Apparently he does," Lurleen said. "I think it's to be close to Lisa. Maybe that's why she plans to give it to him once the estate is settled."

"What's the story with those two?" I asked. "We saw them together at the Blue Cafe looking a lot more like lovers than friends."

"I think they got together in the usual way," Lurleen said. "Brad and Wayne met in college. Wayne meets the second wife years later, and the rest as they say is history."

"That is so cynical, Lurleen," I said.

"I don't mean to be cynical," she said. "It's just that if people stray they often do it with their spouse's best friends. Awful, but true. Anyway, Wayne said that was what happened. Lisa confided in him that Brad was having serial relationships on the side. That was what had broken up Brad's first marriage. Lisa said she was the other woman then and should have known better.

"Wayne comforted her, and then one thing led to another. He told me he knew it was one-sided and thought he never had a real chance with Lisa as long as Brad was in the picture. She'd keep giving Brad one more opportunity to reform.

"Wayne said he got hopeful when he heard about the divorce negotiations. He thought he might get his money and Lisa at the same time. He was pretty devastated when he found out Lisa didn't want the divorce."

"Did Wayne know Brad's first wife, Shannon?" I asked.

"He did and he couldn't imagine why Brad ever married her. He said she must have been really good in bed because otherwise she was a piece of work."

"That's what we heard as well," I said. "Did Wayne elaborate?"

"He said she brought down the Arborough empire in a matter of years. Spent all the money, encouraged Brad to keep her supplied with jewelry and fine houses. The works. And in the divorce she managed to get the old Arborough estate. Not worth much now, but it was then. He said he was surprised when he heard Howie bought it—couldn't imagine why or where he got the money."

"We all saw Shannon at the bar last night. Did Wayne mention seeing her anywhere else—like on the Cape?"

"No. He said he knew Shannon never came to Cape Azul. She hated the beach and Brad's boat and that life style."

"Why so many questions about Shannon, Mason?" I asked.

"Shannon seemed to be pretty angry at Brad and ruthless by all accounts—so I wondered if she'd been staying on the Cape."

"You mean hanging around, waiting for an opportunity to kill him?" Lurleen asked.

"Yes. Of course, that wouldn't have settled her financial issues with him," Mason said. "Better if she tried to get something out of the divorce or out of Lisa."

"And then there was the strange encounter between Lisa and Shannon," I said. "The one Chloe told us about when they hugged each other. Mary sounded as if she couldn't imagine the two of them being in the same room together without a fight—and here they were hugging each other."

"What are you suggesting?" Mason asked. "That the two of them killed Brad to avoid a costly divorce and then agreed to split the proceeds?"

I shrugged. "I don't know what I'm suggesting—just that it was odd encounter." I turned back to Lurleen. "Does Wayne have any chance of making it with Lisa?" I asked.

"I doubt it," Lurleen said. "If anyone's going to make it with Lisa, I'd say it was Scott Huxley."

"That seems to be where things get confusing," I said. "We saw Wayne and Lisa hanging off each other at the Blue Cafe. and we've all heard Scott had a thing with Lisa for years. What's the truth?"

"What's the truth in any of this?" Mason asked. "We're wondering why he hired you, Danny. Scott Huxley doesn't seem that interested in actually finding out who killed Brad."

I jumped in. "If what we heard is true—that his daughter, Erin, might have been having an affair with Brad—then it's possible Scott got angry enough to kill the guy. Maybe the whole investigation is a sham, to get Lisa off his back.'

"What are you actually doing for Scott?' Mason asked.

"He's had me follow up on the men who wanted to hire Brad for a fishing trip."

"Do we know it was a fishing trip they wanted?" Mason asked.

"No That's what the bartender thought—the one who saw them and heard them talking. He didn't know the guys. The men were tourists or said they were. I haven't been able to track them down yet."

"And all we have," I said, "is a bartender who says he saw the men and heard them talking to Brad."

"So Scott's keeping you busy," Mason said, "looking into what might be a red herring."

Danny was quiet for a moment. "You could be right. It's only this bartender who said he saw the guys. I'm getting paid good money to do very little. When I ask if I should look into the diamond and emerald bracelet or into Brad's financial dealings, Scott tells me he's handling that. You think he hired me to make the investigation look legit."

"I wonder," Mason said.

"I've already planned to expand my investigation to include Scott Huxley and his daughter."

"Good," Lurleen said. "It seems Brad liked to give expensive gifts to women he was done with. Maybe he was done with Erin. Did that infuriate *her* enough to commit murder?"

"And what do you know about Brad's younger brother, Howie?" I asked. "He and Brad were estranged after Howie made a move on Shannon, his first wife, before their wedding. Then Howie organized what was supposed to be a memorial service for his brother."

"Scott told me he'd handle all the family connections," Danny said. "He claimed people would talk to him and no one would talk to me about those issues."

"Very convenient," Mason said. "Scott has you chasing down strangers while he quietly handles the rest."

"Maybe that's how he planned it," Danny said, "but that's not how it's going to play out."

"Be careful," Lurleen said. "If Scott was involved in Brad's murder you could be in danger." She put a hand on his arm. "Real danger."

Danny smiled. "Shoe's on the other foot, Lurleen, isn't it? Maybe now you understand how I feel."

Lurleen didn't respond.

"What can he do to me?" Danny said. "He wouldn't risk another murder. He could fire me. If he does that, then he implicates himself."

The doorbell rang, and the pizza was delivered. We let our concerns drop for the evening and enjoyed our dinner.

"Let's regroup in the morning," Mason said. "We've checked out of the hotel and we'll stay here. We're all safe for now and we'll have clearer heads tomorrow."

24

A storm popped up around midnight. It was quite a show. I stood at the window and looked out at the ocean as sheet lightning lit the horizon.

Wait.

Did I see someone walking along the shore? Someone standing and looking at our rental?

I called out to Mason, but by the time he got to the window no one was there.

"You sure you saw someone, Ditie? It's raining so hard."

"I'm not sure, but I think I did. They're gone now. No—look, there, in the water. What do you see?"

Mason squinted into the blackness. We both waited for more lightning. It came, and we saw someone wading through the water. Were they dragging something, looking for something? It was one person, and we couldn't make out if it was male or female. Their head was down. They were wearing a poncho. But yes, they were definitely dragging something. We could make out a rope in their hands.

Mason made a move to get dressed.

"You're not going out there," I said. "You'll either get

struck by lightning or run into trouble with whoever this guy is. He's running his own risk."

As if to make the point, we saw a jagged bolt of lightning strike the sea at the edge of the horizon. The thunder came two seconds later, closer than before.

The storm was coming our way. The man or woman seemed to realize that and climbed out of the water. They pulled some kind of boat out of the water with them.

"It's a kayak," Mason said. "What kind of idiot would take a kayak out in the middle of the night in weather like this?"

"Someone with a death wish?" I asked.

Mason shook his head. "Makes no sense. Unless he was trying to lose something, not find something. Send that boat out to sea."

We watched as the person dragged the boat out of the water along the edge of the ocean and looked up at where we were standing. We hadn't turned on a light, but could he see the shades were no longer drawn?

"He can't see us," Mason said, "but why would he put on this show right in front of our condo? I don't know what this means, but he's obviously up to something."

"You think it's a he," I said.

"He's out there in the wind and rain dragging a boat. That would be tough physically for a woman to do, and frankly a woman would have more sense."

We watched as he pulled the boat out of the water toward our house. We could see more clearly now that it was a kayak. He glanced once more in our direction and then we watched him disappear under the house. All the houses on the Cape were built on steel pilings, so he could enter the lower level where our cars were parked.

"I'm getting my gun," Mason said. "You stay here. I'll wake Danny."

I used a flashlight to look at my cell. No bars. With the storm we had no reception to call 911.

"Be careful, Mason. The doors are locked, so he can't get in. Don't go out looking for this guy, please."

"I won't, Ditie. You stay here. See if you spot him again from this window. Danny and I will be downstairs."

I stayed glued to the window as the storm moved past. Once it was gone, everything looked black outside. The new moon didn't give any light to speak of. I listened for sounds from downstairs and heard nothing. Fifteen minutes later, Mason came upstairs.

"We heard a scraping sound in the carport and then nothing. The storm's over. Danny and I are going to take a look."

"I'm coming downstairs," I said. "Where's Lurleen?"

"Asleep," Mason said, "if you can believe it."

"Should I wake her?"

Mason shook his head no. "You make sure you lock the door after us. Danny and I will check things out."

I looked again at my cell. Still no service. "Couldn't you wait until morning?"

"We need to know if this guy is still around—whoever he is. We'll be careful."

I stood at the front door and locked it after them. I couldn't hear them going down the steps. What I heard seconds later was a sound near the sliding doors by the deck. We'd put a jam in that door. Someone seemed to be tugging at the handle.

I ran into the half bath and grabbed the lid off the toilet—I'd seen someone do that in a movie. It was heavier than I expected and plenty heavy enough to crack a skull—if I could actually lift it high enough to do that. I'm sure it was adrenalin that allowed me to carry it so easily. I stood behind the curtains waiting for someone to break the glass. I knew that was the only way they'd get in.

The sound stopped. Did I hear someone running down the stairs from the deck? Where were Danny and Mason?

Were there voices? A fight?

Five minutes later, I heard knocking on the front door. I left my post.

"It's me, Mason. You can let us in, Ditie."

I kept hold of the toilet lid with one hand and undid the lock with the other—just in case.

Mason and Danny entered and stared at me.

"A toilet lid?" Mason asked, taking it from me. "Hefty. Good thinking, Ditie."

"Any port in a storm," I said, "so to speak."

"You won't need that any more," Mason said. He put it back where it belonged and joined us as Danny started speaking.

"He ran away," Danny said, "but I got a look at him. I think it was Lenny—one of the officers we hired to be present during the wedding."

"One of the people Scott Huxley recommended?" I asked.

"Yeah," Danny said.

"Did you find anything else?"

"We found the kayak," Mason said.

"Anything in it?" I asked.

"We couldn't tell," Mason said. "I figured it could wait until morning."

"Nothing ticking?"

"Whoa," Mason said. "I hadn't considered a bomb. I wonder if you've been hanging around Lurleen too much lately."

"A man was killed, Mason, and I get a feeling some people on Cape Azul don't want us involved with finding his murderer."

"Come on, Danny. Let's make sure there's nothing in that kayak.'

As it turned out there was something in the kayak, something we didn't find until the next day when we pulled it out into the bright sunshine.

25

Lurleen was upset when she woke up the next morning and learned she'd missed all the action. A fan was going full blast in the room, and Lurleen was wearing earplugs—that was the best explanation for why she hadn't heard the commotion.

"How could you have let me sleep through this? Ditie, you of all people, should have known better."

"I'm sorry, Lurleen. All I could think about was that someone might be trying to break into the house."

"And you were going to let me stay asleep while they did that?"

"I would have yelled bloody murder, but it never came to that. Things happened so fast, and then everything stopped. Afterwards, Mason and I went back to bed. Danny stayed up to make sure nothing else happened, and I know if you'd wakened he would have told you."

"Not good enough," Lurleen said. She sipped her coffee. "So what you know is this guy, Lenny, was dragging a kayak near our unit and left it in the carport. Then he ran away before Danny and Mason could stop him."

'That's what we know," I said. "Mason and Danny are

going to check in with Scott Huxley and tell him about the kayak."

"But didn't we decide Scott might not be trustworthy?" Lurleen asked.

"Yes. All they're going to tell Scott at this point is that someone put a kayak in our carport," I said. "Do you want to see it?"

"Are you patronizing me, Ditie? Is this a consolation prize for your letting me sleep through all the action?"

"No. Get dressed and we'll see if the boys missed anything outside or inside that kayak."

Lurleen and I dragged the kayak out to the beach. Anyone who saw us would imagine we were taking it out to sea. The ocean was calm now, but the beach was covered with seaweed. The sun was bright—bright enough for us to see what Mason and Danny couldn't see the night before.

It was a two-man kayak, light enough for us to handle easily and brand new by the look of it. We each picked a side to examine. The brilliant blue paint had not been marred by the storm. We found no dents in the boat, nothing on the outside except the brand name. It had the look of a custom job with darker blue waves painted along the sides. I didn't think it would be hard to trace this kayak, discover who'd purchased it and when. Together we looked on the inside, each of us focused on one seat.

"Look, Ditie," Lurleen said. "What do you think that is?"

Under the seat was a rust-colored patch, the size of my hand. "It could be old blood," I said.

Lurleen nodded. "How did it survive the storm?"

"It was protected—under the seat—if that's what it is."

"It certainly looks like blood," Lurleen said.

We poked around some more but found nothing else.

"So what was Lenny planning to do with this boat?" I asked, "if it was Lenny we saw."

"Did he intend to hide the boat, get rid of it?" Lurleen

asked. Then she answered her own question. "He didn't mean to hide it or send it out to sea—he dropped it in our carport. He knew this was where Danny was staying."

"You might be right, Lurleen. I think he came up the deck —maybe intending to talk to us and then changed his mind. I wonder if he wanted Danny to find the blood, make a connection to Brad's murder."

"But without anyone knowing he was the one who dropped off the kayak," Lurleen said.

"Right."

We finished looking over the boat and then dragged it back under the carport. Lurleen took a photo of it and went to her computer to locate shops on and off Cape Azul where a custom-painted kayak might have been sold. There were half a dozen possibilities. We decided we'd visit each one after Mason and Danny returned.

I used my time to call the kids. We had service for the moment, and I started with Lucie. She'd gotten a cell phone for her last birthday, and I knew she'd be up early.

"Mama," she said. "Hannah and I barely slept last night. Noah's mom didn't mind as long as we didn't wake up the boys. Do you mind?"

"Not at all," I said. "It was stormy here, so we didn't get much sleep either."

"What's going on with the case?" Lucie asked. "Hannah's still asleep, and no one can hear us so you can speak freely."

"Speak freely?" I asked.

"Yes. I know why you were in such a hurry to get rid of us. You wanted to make sure we were safe. Are you safe?"

"We're staying safe," I said.

"What does that mean?" Lucie asked.

I could never put one thing past Lucie. "It means we had a big storm last night and someone dropped off a kayak in our carport."

"Huh?" Lucie asked. "Why?"

"Not sure about that."

"Did it have a body in it?" Lucie asked.

"That's a morbid thought," I said. "No, no body—"

"But what, Mama, what did it have in it? Why would someone do that in the middle of the night, in the middle of a storm? Who left it there? Please tell me everything!"

"You haven't given me much chance to tell you anything, Lucie."

I could imagine Lucie on the other end of the phone trying hard to contain her excitement and probably her anxiety. She stayed quiet and let me speak.

"Here's what we know so far. We found something in the kayak that looked like old blood."

"And you think it might be Brad Arborough's blood—the guy who died," Lucie said.

"Yes, we wonder about that."

"That might mean the killer took the body in the kayak and dumped it out at sea.'

"Yes."

"Thanks, Mama, for not treating me like a baby," Lucie said.

"I'm not convinced you ever were a baby, Lucie, but I've told you that before."

Lucie laughed quietly. "Just a second, Mama. I want to get my notebook. I want to write all this down." She was gone for ten seconds. "I'm back. You don't know who left the kayak?'

"We do have some idea. Danny thought he saw a young man named Lenny running away. He was one of the two police officers at the wedding."

"I remember him," Lucie said. "He was the funny one. He could do magic tricks like Uncle Tommy. He was really nice to us kids. Do you think he's the murderer, Mama?"

"I don't think so. I think maybe he left the kayak, so we'd discover the blood. Maybe to say that was how the body was carried out to sea and floated in so conveniently near the two houses we rented."

"That makes sense," Lucie said. "Uh oh, everyone's stirring. Shall I get Jason?"

"Please, if he's up. Otherwise I'll call him back later."

"I hear him—he's up," Lucie said. "Before I get him, can I tell Hannah what you told me? Is it a secret?"

"You can tell her," I said.

"Good! We're a team like you and Lurleen. We might help you figure this out."

"You might!"

Lucie seemed reluctant to say good bye. "Are you really safe, Mama?"

"I'm safe, Lucie, and always careful."

"Okay then, I'll get Jason."

Jason was on the phone ten seconds later.

"Oh, Mommy, I miss you, but it's really fun here. Noah's mom lets us do whatever we want."

"What does that mean?" I asked.

"She lets us stay up late and play video games. She says it's summer vacation and that means we can have fun."

"I'm glad. Are you getting outside or is it too hot for that?'

"She's taking us to a swimming pool today. When are you coming home?"

"We'll be home in a few days. I miss you like crazy."

"I miss you, Mommy."

"I love you. We'll call you back later on today so you can talk to Mason. Is Noah's mom around?"

Jason left to find her.

Melissa and I spoke for a few minutes. She was happy to have the kids stay for however long we'd be gone. "My boys don't get back for another week and school doesn't start until a week after that. Enjoy yourself."

"You're a godsend," I said. "I don't know how I'll pay you back."

"I love a house full of kids, and it's really fun to have girls here. I don't always know what they're up to behind those closed doors, but I do know where they are."

"I don't always know what they're up to either," I said, "but as long as you know where they are, I'm happy with that. Thanks again."

I got off the phone as Danny and Mason entered our condo.

"Got anything cool to drink?" Danny asked. "It's already roasting out there."

Lurleen poured him a glass of iced tea. "What happened with Scott Huxley?"

"Scott kept asking if maybe we were making too much out of this kayak thing," Danny said. "He told us it sounded as if someone got caught in the storm and just wanted a safe place to leave it. He said he expected someone would be around to collect it today."

"Did you mention you thought it was Lenny who dropped it off?" I asked.

"No," Danny said. "Lenny was in the station when we entered. He didn't speak to us, but he looked worried when we went into Scott's office and closed the door. I asked him to give me a call later, and he said he would. We'll see if he follows through."

"So what have you been doing?" Mason asked. "Going over the kayak with a fine tooth comb? Find anything?"

Mason knew me so well. He knew both of us so well. Lurleen looked ready to burst with information. "Out with it, Lurleen. What'd you find?" he asked.

"Blood," Lurleen said. "We found blood."

The four of us went down to the carport. Danny and Mason examined the spot Lurleen had found.

"Could be blood," Danny said. "What do we do now?"

"We do what we'd do in any other circumstance," Mason said. "We notify the authorities."

"Even if we can't trust the authorities?" Danny asked.

"No choice. If we do nothing, then we may be obstructing the investigation. We'll take a lot of photos, and I'll document the whole encounter on my computer. We'll also take a small sample to send to a lab I know in Tallahassee—just to make sure we get consistent results."

"Do we tell Scott about seeing Lenny run away?" Danny asked.

"You sure it was Lenny?" Mason asked. "It was a dark and stormy night—" he smiled a that—"I couldn't even be sure it was you beside me."

Danny nodded. "You're right. I can't say for sure who ran away."

"Good. I'll call Scott Huxley about the possible blood we found, or do you want the honors?'

"It might have more weight coming from you," Danny said.

Mason made the call. "Scott said he'll be over in an hour. Any chance a starving man could get breakfast while we wait?"

"It's possible," I said. "As long as we can sort through this while we eat."

I made pancakes and bacon. We ate on the deck. A breeze kept things pleasant enough.

"Where are we?" I asked after Mason had poured us all a cup of coffee.

"I'm not sure," Mason said. "Lets start with the players. See if we can get an idea about who the people are that we can trust and those that are more questionable."

"You wish you had your big sheet of paper, don't you, Mason?" I said.

He was an old-fashioned detective and wanted to see everything up on a board.

"You didn't bring a big pad of poster paper with you?" Lurleen asked with a smile.

"I thought I was on vacation," Mason said, "on my honeymoon in fact."

I'll get my computer," Lurleen said. "That's the best we can do for now."

We cleared the table and then moved inside. Voices sometimes carried outside, and we didn't want Scott Huxley or anyone else to hear what we had to say.

Lurleen sat at the head of the dining room table, computer open. "Shall I make a list of good guys and bad guys and questionables?"

"I don't think we know enough even to do that," Mason said. "Let's start with a list of people who had some contact with Brad and might have had a reason to dislike him or want him out of the way."

"That might be a long list," I said. "Shall we start with Scott Huxley and Lisa Arborough?"

Lurleen typed in Scott Huxley. Danny supplied the details he knew.

"He's lived around here for years, and he attended both of Brad's weddings. He has an interest in Lisa from what we've been told and he also has a daughter who may have been involved with Brad shortly before his death."

"A daughter named Erin—not over twenty-one," I added. "Do we think he knew about the relationship?"

"We heard about it in Lost Horizons, but Chloe wouldn't give us her name," Mason said. "I think we need to verify that it was Erin who was involved with Brad."

"I already did that," Lurleen said. "I called Chloe, told her we knew it was Erin having the affair with Brad. She asked us how we knew that, but she didn't deny it. She begged us not to tell Scott."

Danny smiled at Lurleen. "Good work."

Lurleen beamed.

"So it isn't clear if Scott knew or not," I said.

"This is a small community," Danny said. "I'm pretty sure

someone would have told him if he didn't figure it out for himself."

"If he did know, he couldn't have been happy about it," Lurleen said.

"We need to find Erin and talk to her," I said. "I'm certain Chloe knows how we could do that." I turned to Lurleen. "Maybe we need a to-do list."

In the same way Mason loved his big board, I loved my lists.

Lurleen made a new document titled "Assignments." She put interview Erin at the top of the page.

"Back to Scott," Mason said. "You've worked with him, Danny. What's your take on him? Is he a guy who could commit murder to get his daughter away from Brad Arborough?"

26

Danny looked over the list Lurleen had typed on her computer—the one that summarized Scott's history and characteristics. "He does have a quick temper, and he doesn't like things to be out of his control. I've seen that in the office when someone didn't do exactly what he asked them to do. But your question, Mason, is whether I think he might commit murder to protect his daughter from scandal or an ongoing relationship with Brad—that's what you want me to tell you? I honestly don't know."

"It sounds a little over the top to me," Lurleen said. "Brad was a womanizer. We all know that now, and he liked his women young. He was disgusting, but he also seemed to move on quickly. I wonder if Scott wouldn't have handled it some other way. Waiting until it ended. Doing a lock down with his daughter or having it out with Brad—a fight, nothing more."

"I can't imagine a lock-down on a twenty-one year old girl," I said, "one that would work, I mean."

"And a fight might have led to an accidental murder," Danny said.

"We need to speak to Erin," Lurleen said. "Find out what her father knew and when he knew it. Maybe he threatened to kick her out of the house if she was still living at home. We don't really know a thing about her situation."

"A fight between Brad and Scott makes sense," Mason said, "maybe a fight that got out of hand. But the coroner didn't mention any type of bruising on Brad's body, anything that would indicate a fight. Have you actually seen the report, Danny?"

"Nope. Scott promised I could take a look at it; then he said it was misfiled and he'd have to search for it."

"*Mon dieu,*" Lurleen said. "That makes Scott look more guilty than ever."

"Medical Examiner's cases are public records in Florida," Mason said. "I can get us a copy."

"Shall we move on?" I asked. "Tackle Lisa Arborough?"

Lurleen typed up a new heading. "She seems to be a series of contradictions," Lurleen said. "She said she wanted a divorce from Brad, and then she said she didn't want it. She claimed she made all the money in the relationship, but she doesn't have an office on Cape Azul or anywhere nearby. I don't think she's even a licensed real estate agent in Florida."

"What, Lurleen? We've never heard that," I said.

"I did a little checking when I had time on my hands. I was curious. I hadn't seen a single real estate sign with her name on it, so I contacted other agencies that rented or sold houses on Cape Azul. Some had never heard of her. All of them knew Brad Arborough— maybe that's because of his death. The people who had heard of her said she had a couple of houses on the Cape but they didn't think she did any real estate transactions here."

Lurleen smiled at us.

"So, I dug a little more. It's a felony to sell real estate without a license in Florida. You can go to the Department of Business and Professional Regulation's website to verify a license. Lisa doesn't have one."

"But Lisa rented us our wedding houses," I said.

"I know," Lurleen said. "That's one reason I didn't check on her credentials until a day or two ago. I was riding around with the kids and we were imagining what it would be like to live in the prettiest houses we found. That's when I realized I never saw Lisa's name on a single rental sign. How did you get her name, Ditie?"

I had to think about that. "I remember when we were down here last year, Mason and I looked for houses on the beach that might do for a wedding. We took pictures of the ones we liked and wrote down numbers to call."

"So, she was probably just an owner who was willing to rent to you," Lurleen said. "Did she claim she was an agent?"

"Not exactly," I said. "In fact, when I asked about finding a place for you and the kids after the wedding, she told me she was very selective and only handled the larger houses on the Cape."

"*Voila* !" Lurleen said. "She managed to skirt the truth with you. I wonder what other truths she's been skirting."

"All this time we thought she was the one with the money," I said, "and that Brad was using her. It didn't quite make sense —why would she want this man back when he was spending her money on other women? But if Brad was her money source, then it does make sense."

"Don't forget," Mason said, "most people we've talked to said Brad had his own money problems. Big schemes but no money to see them through," Mason said. "Like Wayne Baskin for example. Did Wayne talk to you about that?"

"Yes," Lurleen said, "but he didn't sound furious with Brad. Of course, that was after Brad was dead, so a person wouldn't want to sound too angry at a murdered man after the fact. Still, he described Brad as somebody who liked to play games with other people and their money. Almost like that guy who finally went to jail for the rest of his life over a Ponzi scheme."

"Did Wayne say that's how Brad made his money—from a Ponzi scheme?" I asked.

"No, no. Wayne said Brad never did anything that could get him sent to jail. He always made sure investors understood it was a risk, but he minimized the risk, according to Wayne, and glorified the profit to be had. He'd borrow money from them and never quite pay it back. That was Wayne's situation. Brad kept reassuring him that his investment would pay off big in the end."

"And the investment was in revitalizing Apalachicola?" I asked. "That's what we've heard something about—making it a location for music, theaters, fine dining like Branson, Missouri or Dolly Parton's Dollywood."

"Yes," Lurleen said. "Brad claimed he knew people in the music business who were eager to bring top acts to the town. His vision was that the town would become a mega tourist destination. 'If you can envision it, you can make it happen.' That was his mantra."

"A lot of people in Apalachicola might not have wanted that," I said.

"I wouldn't be so sure," Lurleen said. "According to Wayne a lot of people in the town were eager to jump on board. Of course, Wayne got his information from Brad, so who knows what the truth was."

"We do know Wayne bought it," I said.

"He did. *Comme un bleu*!" Lurleen said.

"Hook, line and sinker," Danny added.

"Comme un bleu?" I asked. "It's an expression for how the French cook fish, isn't it?"

"I taught Lurleen that phrase a week ago—came in handy today," Danny said. He'd minored in French in college and French was one of many similar interests he and Lurleen shared. Lurleen had her own unique version of the French language, but she was happy to have Danny teach her legitimate phrases.

"Nice nautical analogy, Lurleen," Mason said, "given our current situation. Lurleen smiled broadly. It wasn't too often she got a direct compliment from Mason.

"Actually, it isn't nautical at all," Danny said. "It comes from the blue uniforms that rookies wore in the French military."

Lurleen cut her eyes at Danny.

"Still a good term," Mason said. "A lot of people seemed to believe whatever Brad told them—hook, line and sinker."

"Wayne must have been furious with Brad for conning him," I said.

"He didn't act like it," Lurleen said. "He didn't even act like Brad had actually conned him. He said Brad made it all seem fun and exciting in the beginning. Wayne had the money to invest at the time, so he gave it to him. After a few years Wayne could see the big investment in Apalachicola wasn't happening. Brad told him it would happen—soon—and all Brad needed was a little more money from Wayne. Wayne gave it to him."

"You're kidding," I said.

"No," Lurleen said. "According to Wayne, Brad was a super salesman. Boatloads of charisma—to continue our nautical theme. Wayne was happy to go along for one more ride—this time as a loan. Then he began to have his own financial troubles and needed to get that money paid back. Brad had assured him that any time he wanted to get it he could. But, of course, that didn't happen. So Wayne came down here to get the loan money back."

"And see where he could get with Lisa," I said.

"Yes, He saw coming down here both a risk and an opportunity,' Lurleen said. "You know that's how investment bankers like to talk. He thought he might forfeit all the money he'd loaned Brad—say if Brad declared bankruptcy—but he also saw it as an opportunity with Lisa. As I told you, Wayne's a softie, a romantic. He told himself he might have a chance with Lisa and he shouldn't pass it up. We've seen how Lisa flirts with everyone, but Wayne couldn't see that. Poor man."

"I need to take a breath here," I said. "What you seem to be describing is how one con man—Brad—married another con man—or in this case, woman—Lisa."

"That's about the size of it," Lurleen said. "What Wayne told me was that however much Brad understood about conning people, what he didn't seem to realize was how upset people got when they figured out they'd been conned. For Brad, it all seemed like a big joke."

"For someone else it obviously wasn't," Mason said.

27

Lurleen stood and stretched; then she sat down again at the table in front of the computer.

"You're right, Mason," she said. "Someone didn't find Brad's antics amusing. Or forgivable. I asked Wayne if he knew who else had invested in Brad's scheme to make Apalachicola another Dollywood. He rattled off a list of likely investors—Scott Huxley was on the list. He also put Mary from Lost Horizons on the likely investor list if you can imagine that."

"Mary wanted Apalachicola to expand in that way?" I asked. 'I would have thought Mary might have wanted Apalachicola to remain a sleepy little river town."

Lurleen shrugged. "We should ask her about that. Maybe she wanted to sell more of her jewelry."

"Who else was on the list?" Mason asked.

"Shannon, Brad's first wife. Oh, and Brad's younger brother, Howie."

"Howie?" I asked. "When we saw Howie he didn't look like a wealthy man. I wondered how he paid the bar tab from the memorial service."

"Wayne said looks can be deceiving. I guess the possibility of a windfall from the divorce brought everyone back together," Lurleen said. "Wayne told me that Howie was seen in town when the old mansion was auctioned off," Lurleen said. "Now we know who sold it—Shannon—and we heard at the memorial service it was Howie who bought it."

"I wondered how Howie had the money to buy that place," Danny said, "even if Shannon sold it cheap."

"Or why he'd want to," Lurleen said. "It would cost a fortune to fix it up if it even could be fixed up."

"And it was clear from the memorial service," I said, "that Howie wasn't a popular man in Apalachicola after what his brother did, so why would he want to stick around?"

"We're left with the question of how Howie suddenly had the money to buy the house," Mason said, "and why he'd want to do that in the first place."

"So many questions," I said.

"So many suspects," Lurleen added. "Howie, who lives like a beach bum, suddenly has the money to buy his own home and wants to do that for some reason. Lisa, supposedly the money maker, perhaps really lived off Brad's money and wanted all of it.

And Shannon? Wild card Shannon. She wanted money from Brad—not the best motive for murdering him. What did she think she could get from this second divorce?"

"And then there's our prime suspect Scott Huxley," Danny said. "Lost his money to Brad and maybe his daughter. Would that be enough to push him over the edge and make him want to kill the guy?"

"Sounds possible to me," Mason said. "At least he might want to beat him up, and if it was a fight gone wrong—that could explain the rest. Maybe Scott didn't plan to kill Brad, and then when he did, he had to get rid of the body."

"Right," Danny said. "It would be easy enough to unmoor Brad's boat, send it out to sea, empty. Then all he had to do was carry Brad's body in his kayak, late at night, drop it in the ocean and let it turn up with the tide."

"Would he want the body found?" I asked. "And why near the houses?"

"Good questions," Mason said. "I can imagine he'd want the body found, particularly if he was handling the investigation. It could look as if Brad was drunk, bumped his head, and fell overboard. That would clear the way for Lisa to claim the estate, no questions asked, no messy divorce to eat up money. But Scott would know if Brad was dead before he went into the Gulf his lungs wouldn't fill with water—any Medical Examiner would discover that."

"And that's what the Medical Examiner did find," Danny said. "Scott admitted that, even though he claimed he'd mislaid the report."

"As to why Scott would want the body found near the houses Brad and Lisa owned, that's harder to understand," Mason said. "Maybe he didn't want that—maybe it was how the tide brought the body in."

"You don't believe in coincidences, Mason," I said.

"You're right. I don't. That part of the story I can't explain. I think we need to get a few facts straight. Let's see if it is blood we found on the kayak, and let's get hold of the Medical Examiner's report."

"Lurleen and I will try to track down who bought the kayak," I said. "And while we're at it, we'll try to have another talk with Mary and Chloe. I can't believe Mary would be in favor of turning Apalachicola into what Brad had in mind, and I'm pretty sure Chloe can tell us how to contact Erin."

"You can try to track down who bought the kayak," Mason said, "and you can talk to Mary. I'm not sure you should try to talk to Erin—she's a suspect."

"We'll see how it works out, Mason," Lurleen said. "I promise we won't talk to Erin unless it's in a public place. All we really want to know is whether or not she was still in a relationship with Brad when he died and if her father knew about it."

"Don't press her," Mason said, "and don't accuse her of anything. For you, Danny, it needs to seem like business as usual. If you get a chance to talk with Lenny take it, but only if you can do it when Scott's not around."

"I'm not sure that will be a problem. Scott's spending a lot of time with Lisa from what I can tell."

"Where?" I asked. "Where are they spending their time?"

"On Brad's boat. There's a lot of privacy there."

28

"Wait a second," I said, "let me get this straight. Scott and Lisa are hanging out on Brad's boat—a boat that is potentially a crime scene."

"Now, it seems less likely it's a crime scene," Danny said. "Lenny dropped off that kayak perhaps to tell us the murder took place somewhere else and the kayak was used to drop Brad's body into the sea."

Mason nodded. "Even so, there must be other private places on Cape Azul."

"It's a small peninsula," Lurleen said, "where everyone keeps an eye on what's going on."

"Did Scott admit to you, Danny, that he and Lisa were involved with one another?" I asked.

"No," Danny said. "Scott claims he's helping Lisa deal with her grief and providing a quiet place for her in which to do that."

"You make it sound like it's *his* boat," I said.

"That's the most surprising part of the whole story," Danny said. "Scott claims Brad gave him the boat for a song — a deal Scott couldn't resist."

"Lisa told us it was Brad's prize possession," Lurleen said, "and that she planned to let Brad keep it in the settlement."

"According to Scott," Danny said, "Brad became a different man in the last days before he died—a kinder and more generous one."

"We certainly didn't see that side of him," Lurleen said, "when he tried to rent the wedding house out from under us."

"Is anyone giving us a straight story?" I asked. "Scott had every reason to want Brad dead—from an affair with his daughter to schemes that cost him his retirement savings and his wife. Now he's telling Danny that Brad had some great personality transformation before he died. How does that make sense?"

"It doesn't make sense," Lurleen said, "unless you remember how well Brad could con people. Some people continued to buy what he was selling. Maybe Scott was one of them. I know Wayne wasn't as angry at Brad as I expected him to be."

She stopped talking for a moment. "Wait a sec," she said. "I just remembered something Wayne told me. Right before I left him, he said that everything might work out for him. Brad offered to give Wayne some money as a down payment on what Brad owed him. When Wayne asked where he got the money, he said he was finishing up a business deal and was cleaning house for a brighter future."

"Cleaning house? A brighter future?" I asked. "Maybe Brad actually was changing in the days before his death."

"But that doesn't explain where he suddenly got his hands on a lot of cash," Mason said.

"No, it doesn't," I said.

"So what was Brad up to in the days before he died?" Danny asked. "Was he blackmailing someone? Is that where he got his money?"

"How does that match up with cleaning house and a brighter future?" I asked.

"Not very well," Danny said.

"And it doesn't explain why he was making the divorce so difficult," I said, "according to Lisa."

Lurleen spoke up. "Wayne said he thought Brad had money stashed away in assets that couldn't be touched even if he went bankrupt."

"How do you pull that off, Lurleen?" I asked.

"You put a lot of assets in your wife's name," she said, "and do a few legal maneuverings that any accountant and a good lawyer could set up."

"Ah, so that might explain why a divorce could get sticky," I said.

"Yes. Lisa could walk away with a lot of money," Lurleen said. "And yet she says the divorce was Brad's idea."

"A lot of how Lisa acts and what she says doesn't match up." This was from Danny. "She talks about how she wanted a reconciliation with Brad and then makes a play for every man within range."

Some alarm bell must have gone off in Lurleen's head. "What are you talking about, Danny? Has she made a play for you?"

Danny squirmed. "Not directly, but she managed to catch me in the bar near the entrance to the park—the Blue Cafe. I was talking to the bartender who said he'd seen the two men with Brad. Lisa was super friendly with the bartender and acted so surprised to see me there when she was the one who suggested I talk to the guy. She offered to buy me a drink, and when I said I didn't drink on the job, she laughed at that. She put a hand on my arm and said I was the first PI she'd ever met who didn't drink on the job. She kept pressing, like—who would know, who would it hurt?"

I could see steam rising from Lurleen's auburn curls—I swear I could—like an ominous sunset.

"She *was* making a play for you," Lurleen said.

"Honestly, Lurleen," Danny said, "I think Lisa makes a

play for every guy she meets and probably always with an ulterior motive." He hugged Lurleen. "You'd think she'd realize she didn't have a chance with me."

"Brad certainly knew how to pick 'em," I said, "wives, I mean. His first wife, Shannon, flirts with his younger brother before the wedding, and Lisa flirts with everyone after she marries him."

"Don't forget, Brad did the same thing," Mason said, "birds of a feather."

I nodded. "You're right. I wonder what made their house of cards fall down in the end. Why would either of them want a divorce when they could both do exactly what they wanted anyway?"

"Is it possible one of them wanted a different kind of life?" Lurleen asked. "Maybe one of them actually fell in love with somebody else."

"Are you thinking what I'm thinking, Lurleen?" I asked, "that maybe Brad actually fell in love with Erin and wanted more than just a fling."

"That's an interesting theory," Mason said.

"It could even explain Brad's change in personality," I said, "including his desire to clean house, make amends—especially to Scott."

"I wonder," Lurleen said. "She's twenty-one years old when Brad had to be closing in on fifty. People said he liked his women young but that nothing lasted. Then there's the bracelet, meant for someone, although we don't know who was supposed to get that gift. Was it a parting gift or a gift to signify undying love? If it was for Erin and if Brad intended to marry her after the divorce, a lot of people might be upset about that."

Lurleen looked at her watch. "Let's get going. It's nearly lunch time and all we've done is talk. Let's see what we can figure out before dinner."

29

Scott Huxley knocked on our door as we were getting ready to leave our condo.

"Going somewhere?" he asked. "Glad I caught you. I'm here about the kayak."

"Downstairs," Mason said, "in the carport—I'll show you."

I hoped Mason would find a way to get his own scraping of the questionable blood.

Lenny was with Scott, and he barely acknowledged us. Lurleen and I waited upstairs. I took the time to check in with the kids. Melissa told me they were off to the swimming pool in a few minutes, enjoying the last of summer vacation.

I hung up glad to know the kids were having fun.

"I have service at the moment," I said to Lurleen. "Should I call Lost Horizons and see if we can talk to Chloe? See if she'll give us a lead on Erin?"

"It's worth a try," Lurleen said.

Chloe was the one who answered the phone. She didn't sound altogether happy to hear from me. I asked her if any customers were there.

"No, just me. Shall I get my grandmother for you?"

"Not right now. You can tell her we plan to stop by later if she'll be around. You can say my friend Lurleen wants to see her jewelry."

"But that isn't why you're coming, is it?" Chloe said.

"Not entirely."

"What do you want from us, Dr. Brown? We don't know anything about Brad's death."

"But you do know where Erin is. We need to speak with her about her relationship with Brad."

Chloe lowered her voice. "Erin is very upset about Brad's death—she really thought he might marry her. I told her that was never going to happen, but of course she wouldn't listen to me."

"Can you tell us how to find Erin?"

Silence.

"Please, Chloe. We need her help."

"She loved Brad. She didn't kill him."

"You may be right," I said, "but someone killed him. If we can talk to Erin, maybe we can start to figure out who that person is."

"I'll ask her. If she's willing to talk to you she'll be in the store this afternoon. What time are you coming?"

"Let's say three."

We hung up. Mason and Danny joined us.

"Scott's taken the kayak. I showed him what looked like dried blood. I scraped a little so he could see what I was talking about. Then I managed to give it to Danny when Scott wasn't looking. You still have it Danny?"

"Sure," Danny said. He showed us a baggie with the flakes inside. "Hope it will be enough to identify it."

"We'll know if it's blood and probably get a blood type," Mason said. "I doubt that we can get more information than that. DNA testing is expensive, and we'd have to have a good reason to ask for it. But at least the lab will have a sample if we need it in the future."

Mason gave me a long look, but I couldn't read his expression.

"We're heading for Tallahassee," he said. "We'll drop off the sample and see if we can get a copy of the coroner's findings. We'll be gone most of the day, and I don't want you two going into any questionable places while we're gone and can't help."

"When you can't rescue us, you mean," Lurleen said.

"That's exactly what I mean," Mason said.

"We're going to see if we can track down the person who bought the kayak," I said, "and we're going to talk to Mary and Chloe again at Lost Horizons. We'll see if we can get in touch with Erin."

"That's store is a rabbit warren," Mason said.

"We promise not to get kidnapped," Lurleen said.

"Not funny," Danny said.

Mason looked uneasy.

"We'll be careful, Mason," I said. "We won't go into any dark places. We'll stay among people and we'll be home by five."

"Good. I can't say when we'll get home, but I'll leave you a message—I'll text or email you if I can't get through by phone."

We watched Mason and Danny leave.

"He spooked me, Lurleen," I said. "I have to admit that. We can't do anything the least bit dangerous."

"I know. We won't."

How many times had Lurleen said that to me? How many times had she done something just a little off kilter—just enough to get us in hot water?'

I looked at her.

"We have two missions today," Lurleen said. "Check out

some shops selling kayaks and talk to Erin and Mary at Lost Horizons."

"If Erin will talk to us."

"Yes. That's it. That's all," Lurleen said.

Of course that wasn't all. With Lurleen that was never all.

30

Lurleen and I started our search for the purchaser of the blue kayak by heading north on Cape Azul. We began with the small shop located next to the Blue Cafe. We'd work our way down Cape Azul and then move farther afield.

No luck at Joe's Kayak and Surfboard shop.

"Haven't sold anything like what you're describing," Joe said. "Don't even carry something like that. You said it's a shiny sky blue one with some waves along the side? We do some custom work, but we didn't do that one."

"Thanks," Lurleen said.

"Look around. My stuff is custom made. We buy name brands but then we add our own touches. You might see something you like."

Two more shops on the Cape said they didn't carry the brand of kayak we described. We headed to the town that was in the opposite direction of Apalachicola but not far. There was a big store that carried all kinds of boating and outdoor equipment, including the brand of kayak we were looking for. We had a different problem in that store.

The salesman called his manager when it seemed clear we wanted information and didn't intend to buy anything.

"Why'd you say you needed to know who might have bought a blue kayak?" the manager asked.

"We liked the look of it," Lurleen said, doing her best to sound like an interested customer. "We saw someone in the ocean with it on Cape Azul and thought if we could know who bought it, we might contact him, see what he thought of it."

Oh, boy, even I knew this wasn't going to fly. Lurleen tried to pour on the charm but this guy wasn't having any.

"You with the police or something?"

"No," Lurleen said, trying to look aghast.

"I didn't think so. No one sees our records except the police."

We moved on.

"No one from a big store is going to give us the information we want," I said, "and whoever bought the kayak might have bought it from a big store where he wouldn't be noticed."

"Except that the kayak had custom work done," Lurleen said, " and a big store wouldn't do that."

"No, and if someone bought it intent on doing something illegal," I said, "they'd never have custom work done."

"Something illegal like dump a body?" Lurleen asked.

"Yes," I said, "so that means the kayak wasn't bought for that purpose."

"Right," Lurleen said, "and it looks new, no scratches, nothing. It looked expensive, and add to that the custom painting on the side—someone spent money on this, and it wasn't so they could dump a body out at sea."

"Too bad," I said."The person who knows about the kayak is Lenny, and we can't approach him."

'We can't?" Lurleen said, flicking back her auburn hair and giving me her most ravishing smile.

"No, Lurleen, we can't. But we can check out small kayak

shops on our way to Apalachicola, and you can use your charm to get information."

We found two more small stores selling kayaks, one just outside Apalachicola and one off Main Street. The first shop offered nothing despite Lurleen's request. Oh, the guy was happy to offer Lurleen whatever she wanted, but he hadn't sold a light blue kayak recently or ever for that matter.

The second store was down the street from Lost Horizons, and it was there we struck gold. Lurleen claimed she was interested in buying a kayak. She'd seen one just like what she wanted on Cape Azul. Had the owner sold a blue one with hand-painted waves along the side?

"Sounds familiar, let me check." He went to his computer, pulled up recent sales and found one.

"Found it. It's the brand you're looking for. I can order it for you with or without the custom work."

At that precise moment, Lurleen doubled over with a coughing fit. "Water," she gasped. "Could I get a glass of water?"

"Sure, sure," the guy said.

"Oldest trick in the book," Lurleen said as he headed for his bottled water supply.

She scooted around the desk, refreshed the computer and found what she wanted. She was back in place when he returned. He handed her a bottle of water.

"Thanks so much," she said, gulping down some of it. "How much do I owe you for the water?"

"On the house," the man said. "You want me to order you that kayak? It's the best we have."

"Let me think about it and get back to you. I never make impulsive decisions."

I had a hard time keeping a straight face.

"Hmm," the guy said. I think he realized something wasn't right. "The bottle of water is two bucks."

Lurleen handed over the money and we left the store.

"So?' I asked. "Who bought the kayak?"

"Erin Huxley," Lurleen said. "Looks as if she used her dad's credit card—his name was on the card."

"You're kidding,"

"Would I kid about something like that?" Lurleen asked.

"No," I said. "Erin buys a kayak, uses her father's credit card to do it. All easily traceable. So, either the kayak wasn't meant to be used to drop a dead body in the ocean or Erin isn't a very smart girl."

"And why would Erin want to kill the man she loves?" Lurleen asked.

Then she answered her own question. "What if that man —Brad— had already moved on? We know Brad liked to give expensive gifts to women when he was done with them—maybe the bracelet was meant as a parting gift for Erin."

"We are through the looking glass," I said. "Let's see if Erin is willing to talk to us."

31

We walked down the street to Lost Horizons. I glanced at my watch—three p.m.—we were precisely on time. We entered the dimly lit store and took a moment to adjust to the lighting before we moved anywhere.

I loved the store with its dark wood cabinets and paintings covering every bit of wall space. It wasn't glitzy like the newer tourist stores in Apalachicola, and I was glad Mason and I had spent some money in it recently. I couldn't imagine that Mary, the owner, would want to see Apalachicola turned on its head, assuming Brad Arborough could bring his dreams to fruition. It didn't make sense she'd invested in his schemes.

Chloe came up to us before I registered she was there.

"Hi," she said quietly. "I'd rather not have Grandma know you're here while you're speaking to Erin. She's willing to talk to you, and she's waiting around the corner."

"Thanks, Chloe," I said.

We left the store as quietly as we'd entered. Chloe must has silenced the bell above the door that normally announced visitors, in case Mary was in the back.

We found Erin around the side of the building, seated in a

shady spot on a bench. Beyond her we could see the door to another older store and beyond that the river.

" You must be Erin," Lurleen said. "I can see you chose a private spot for us to talk—not much foot traffic around here."

Erin stood, shook our hands and then sat down again. She was a beautiful girl who looked fresh, younger even than her years. I'm not sure what I expected, maybe someone wearing too much make-up. I didn't think she was wearing any. She was a tall girl, close to Lurleen's six feet, with her blond hair pulled back in a pony tail. She wore short shorts that showed her long legs and a t-shirt that revealed nothing other than an advertisement that said ApalachiCOLA with a can in Coca Cola colors.

"You wanted to talk to me?" she said.

"Yes," I said. "Chloe told you about us?"

"Chloe told me you wanted to talk about Brad. I figured you'd probably talk to my dad if I didn't speak to you, and I don't want that, so ask your questions."

"I'll start with the obvious one," Lurleen said. "What was your relationship with Brad?"

"We dated some," Erin said and fidgeted with her pony tail. "It was no big deal—really just a fling. We both knew that. But my dad didn't. When he found out, he went ballistic. He made me stay away from him, and Brad didn't even fight for me. Brad said he had to keep in my dad's good graces."

"That must have hurt," Lurleen said.

"It stung me, but I was already about over him anyway. I mean a guy that old—sure it's flattering when a guy like Brad wants you and wants to shower you with presents, but only for a little while. He couldn't keep up, not really. So no big deal."

She stopped herself.

I wondered if she stopped because she was putting on a performance and needed to catch her breath.

"I'm not saying it isn't horrible he died," she finally continued. "That was horrible, but we were done with each other long before that happened."

"How long before he died?" Lurleen asked.

"What? Why do you want to know? You think I had something to do with his death?"

"You said your father was furious when he found out about the two of you," Lurleen said.

Erin stood. "You're accusing my father of killing my ex-boyfriend? Boy, have you been watching too many dark TV serials. We don't do that kind of stuff here."

"Someone does that kind of stuff," Lurleen said. "You know Brad was murdered, don't you? He didn't just fall off his boat drunk even though the murderer hoped we'd think that."

"My dad told me it was murder," Erin said. "You think he'd have told me that if he were the one who killed Brad? Dad was relieved when I said it was over. He told me all his dealings with Brad were over as well. We both moved on."

"When?" Lurleen asked. "When did you both move on?"

"Weeks before Brad died. If that's all you want from me, I'm busy. I need to go."

She stood as if to leave.

"There is something else," I said. "You bought a kayak not long ago, a blue kayak."

"So what?"

"So someone left that kayak under our rental unit at night during a storm," Lurleen said, "and we found what looked like dried blood underneath one of the seats."

"What?" Erin's face drained of color, and she sat down hard on the bench.

Lurleen sat down beside her. I continued to stand in front of them both, the sun hot on my back. No one was near us.

"Blood?" Erin asked. "Whose blood?"

"We don't know yet," Lurleen said.

"It's not Brad's. It can't be Brad's." She started to cry.

Lurleen put a hand on Erin's arm. "What's going on?"

"At first Dad told me that Brad fell off the boat and drowned—that it was an accident. He said Brad was drunk at the time. All I could think was that if I'd been there, it wouldn't have happened. But Dad wouldn't let me go anywhere near Brad once he knew we were involved. Brad said I had to be patient until everything was arranged. He said I had to obey my father for a while. Brad told me he had something important to do and once he'd taken care of that, we'd get to announce our relationship to the world. We'd get married."

Now, Erin's tears became a torrent. We waited until they stopped.

"So you were still involved with him when he died," I said.

"Yes. After Brad died, my father said I wasn't to tell anyone anything. He said it would complicate his investigation, and that's when I realized Brad had been murdered."

She rubbed her eyes, and I was afraid there might be a fresh flood of tears. Lurleen put a hand on Erin's arm to steady her.

"All I have left of him is the fact he loved me. He swore he loved me in a way he'd never loved anyone before." She wiped her eyes. "He said soon it would be clear sailing. He said even my dad wouldn't object to us at that point."

"But your dad did object," Lurleen said.

"Dad is such a hypocrite. He said Brad was still married, and I couldn't run around with a married man. I saw my dad with Lisa Arborough dozens of times. She was married and she wasn't the first married woman he'd been involved with."

"He didn't object because of the age difference?" I ask.

'Oh yeah, he and everyone else was against us because of that."

"Who is everyone else?" I asked.

"Chloe for one. She said I was throwing my life away on someone who wasn't worth it. Lisa said the same thing. And then Howie shows up—Brad's younger brother. He told me he was a better catch than Brad, that he had more going for him. Even Shannon, Brad's first wife, told me I was making a big mistake."

"You really do mean everyone," Lurleen said. "Any old boyfriends in the picture who might have gotten upset about the relationship?"

"Just Lenny. Lenny and I grew up together, but Lenny had to know we'd never end up as a couple. He was like a brother to me. Nothing more, and he never would be."

"Did he accept that?" I asked.

"I don't know. He said he'd do everything he could to keep Brad away from me, but Lenny is all talk. He couldn't hurt a fly—I can't imagine how he does his work as a police officer. Not much happens on Cape Azul, maybe that's how."

"This may be a hard one for you to hear," I said, "but lots of people have told us Brad went from one relationship to the next and that he liked his women young. They also said before he left a woman, he'd give her an expensive piece of jewelry to let her down gently."

Erin shook her head vehemently from side to side. "Our relationship was different. Brad told me that. He said he'd never felt about anyone the way he felt about me. He'd never loved anyone before me with his whole heart."

"So he didn't give you any jewelry before he died?" Lurleen asked.

"So what if he did—it wasn't the same. He showed me a bracelet and said he'd give it to me as soon as he cleared up his one last problem—it would be a wedding present."

"A diamond and emerald bracelet?" I asked.

"How did you know?"

"We found it under a pillow in the master bedroom at the big house," Lurleen said, "all wrapped up like a gift. Is that what he promised you?"

I found a picture of the bracelet on my cell and showed it to her.

Erin nodded. "But you don't know what you're talking about. Brad loved me. For the first time in his life, he was really in love."

Lurleen sat back on the bench. "Let's say you're right about that. You've listed a half dozen people who hated the idea of you and Brad together. Could one of them have hated it enough to kill Brad?"

Erin sat up straight, hands underneath her and turned away. For a few seconds she stared out at the river and said nothing.

"What is it, Erin?" I asked.

"It's the blood in the kayak," she said and started to weep. No sobs, just tears streaming down her face. "Was it Brad's blood?"

"We don't know yet," I said. "We may never know that. Did someone ask you to buy that kayak for them?"

Erin nodded. "My dad said I should buy it, and that I could use his credit card. Dad apologized for not being a better dad, not spending more time for me. He thought that was why I was finding comfort in an older man like Brad. He didn't understand Brad or our love.

"But I bought the kayak anyway and nothing changed. Brad died before my dad and I could take it out together even once." She looked at both of us. "My dad didn't kill Brad. He's not that kind of guy." She stood up again. "I gotta go. Can't you leave us in peace? Brad's dead. Why does it matter who killed him? This is our private business—not yours."

32

We watched Erin walk away.

"She's scared to death her father might have murdered Brad," I said.

"She is—*la pauvre chérie.* Can you remember what it was like to be in love with someone when you were twenty-one?"

"I can," I said. "I thought I was so mature at the time—I was in med school, and I was certain I'd found the man I was going to marry. I knew Phil was the one. I imagined we'd be two doctors side by side saving the world together. You and everyone else told me he wasn't right for me, but I couldn't see it. At that age, when you meet a man who says he loves you—a man you love back—it's your world."

"Yes," Lurleen said. "I saw enough of my mother's boyfriends to know that wasn't the case. I saw how they used her, got her hooked on drugs. They put me off men for years. But for people like you and maybe Erin, those first loves feel so powerful."

"People like me?" I asked.

"You know, people who trust the world and the men in it. You adored your father. You thought he hung the moon. And when you met Phil you gave him all the attributes your dad had. I never knew my father so I couldn't do that. I saw my mom with men who took advantage of her and hurt her. It took me a long time to trust Danny enough to let him into my life."

"I know what you mean, Lurleen," I said. "It took me years to get over Phil. I'd probably still be single if Mason hadn't turned up when he did and been the man he is."

"Out of a terrible tragedy, something good can come," Lurleen said. "Out of Ellie's death Mason found you and you rescued two children—children who needed you."

"I needed them. I just didn't know it at the time," I said.

"That's what my aunt Theresa always told me—that she needed me more than I needed her. We were strangers when we met, and she was the one who took me in when my mom died. But Theresa always said I was the one who rescued her. She'd lost her own daughter to drugs, and she said I brought joy back into her life."

Lurleen had tears in her eyes. "We had some wonderful years together before she died."

"I have no doubt about that, Lurleen."

Lurleen stood and looked off into the distance shaking her head.

"I hope Scott had nothing to do with Brad's death so Erin can heal," she said, "and she *will* heal. She can have a bright future, but if Scott killed Brad, I don't know how she'll get over it."

"It doesn't really make sense," I said. "If Scott meant to use that kayak to drag a body out to sea, he'd never have had Erin buy it. I wonder if he meant to use it just the way he said—as a chance for the two of them to be together."

"That would mean," Lurleen said, "that either something went wrong suddenly like a fight with Brad that turned deadly—or that someone else used that kayak."

"Yes," I said. "I think it's time we have a talk with Lenny.

Erin may have thought of them as brother and sister, but it's not clear Lenny thought that."

"But," Lurleen said, "if he were involved with Brad's death, he'd never have brought the kayak to our house."

"Unless he wanted to throw suspicion in someone else's direction," I said. "We need to see what he has to say for himself."

"We're here," Lurleen said, pointing to the weathered sign for Lost Horizons. "Shouldn't we talk to Mary and Chloe first?"

"Absolutely."

We entered the store and once again had to let our eyes adjust to the darkness inside.

Mary was standing at the counter helping a customer. Chloe was at the cash register finishing up a sale. Lurleen and I walked around the store looking at the paintings on the wall. Most of them were water colors. A few were by the artist whose paintings I'd purchased.

Lurleen spotted the picture before I did. It showed a large rambling house, almost a plantation home but not nearly as tidy or as grand.

"Doesn't that look like—?" she said to me.

"The Arborough house," I finished for her. "In its finest days."

"Yes. It was a lovely house wasn't it?" Mary came over to stand beside us. "It was a showpiece in Apalachicola in the early 1900s, an anchor," she said, "before it became an albatross."

"When and why did that happen?" I asked.

"Styles change, I suppose," she said. "Everything in Apalachicola changes on a dime."

"I've noticed that," I said. "Every year I come down here, and every year I don't know what to expect."

"Exactly," Mary said. "Of course, some of us hang around forever—like me and now Dawn with her bookstore.

But other things change. It's a river town and I guess we should expect it to be constantly changing like the water that flows into Apalachicola Bay."

"I really like this picture," Lurleen said. "It tells us where this town came from. How much is it?"

"That one's not for sale."

"Too bad. Who painted it? I can't make out the name."

"I painted it," Mary said, "back in the day when I thought I might live in that house."

"There's a story," Lurleen said. "Will you tell us what you mean?"

Mary looked over at Chloe to see how busy she was. There was only one other customer in the shop and Chloe was bringing out rings for her to try on.

"It was after WWII," Mary said, "when the boys were returning. Willie Arborough came back from the war—single and looking so handsome in his uniform. I was a teenager at the time—a little younger than Chloe is now. I didn't know Will was sowing his wild oats before he settled down. I believed him when he said he loved me. He showed me the big house, said one day we'd live there. Next thing I knew, he was engaged to be married to a young woman—an Alabama socialite that his family approved of—and that was the end of me. My mother scolded me for being so naive. She said I didn't know my place—how could I ever have imagined the likes of me getting married to the likes of him?"

"What a sad story!" Lurleen said.

"It wasn't sad in the end. I met a man who really did love me. He was a fisherman, made good money, let me live my life as an artist. He came up in the world as I watched the Arboroughs sink lower and lower. I think their own greed got the best of them. For a while they were as famous around here as John Gorrie was for his invention of the ice machine. Their inventions had more to do with fishing—streamlining things, but like a lot of wealthy people, nothing was ever enough."

She stopped as if all her talking had exhausted her. We waited.

"Willie married, as I said, and had a boy, Arthur. He followed the family tradition and married a socialite, this one from Atlanta. They had two children—Brad and Howie. You know the rest."

"Not really as much as we'd like to know," I said.

"You've heard the adage," she said, "the first generation makes the money. The second generation spends it, and the third generation is left with nothing. That's basically what happened to the Arboroughs. Brad and Howie were left without much except the desire to be wealthy. They were like barkers at a carnival—all talk and con."

"You must have been upset when you heard about Erin getting involved with Brad," Lurleen said. "Deja vu all over again."

"Upset? She isn't my relative. Now, if it had been Chloe—that's a different story. But Chloe's best friend? Well, I was sorry about it, but I also knew you don't tell a twenty-one year old anything. She had to find out for herself, and fortunately she did."

"That's not what she told us," I said. "She seemed very upset over Brad's death. She claimed they were going to be married."

Mary snorted. "Erin wanted to save face, that's what I think. Chloe told me she'd discovered Brad in bed with another woman, and Erin was furious."

"You know the name of this other woman?" I asked.

"Didn't ask," Mary said. "None of my business."

"One more question," I said. "We heard you invested in Brad's scheme to turn Apalachicola into a second Dollywood."

"I'd never do that," Mary said.

"So you didn't invest or lose any money because of Brad?" Lurleen asked.

"That isn't what Brad described to me," Mary said. "Brad told me he wanted to make the town a cultural center, a place where artists could gather like Asheville, North Carolina. That sounded good to me, so yes I invested. I lost every penny, and I suppose that's just as well if that wasn't what Brad really had in mind."

We thanked Mary, and she disappeared into her back room.

"Brad could sell snow to Eskimos," Lurleen whispered.

"Apparently," I said. "I wish we'd been able to hear how he did it."

"Oh, Ditie, you and I would have been swept right along with the rest. Well, maybe not you, but I'm a romantic. I would have believed in anything he was selling."

33

We watched Mary return to her inner sanctum. Chloe was still busy with her customer, so we stepped outside.

"I can't tell the lies from the truth," I said. "Erin claims she and Brad were going to marry, and Mary tells us that was all nonsense—that Erin knew Brad had moved on to a new affair."

"Just a typical murder investigation if you ask me," Lurleen said, "full of untruths. Erin would be plenty strong enough to load a dead body into a kayak and drop it out at sea. Maybe if she spotted the other woman in one of the two houses, she'd want Brad's body to wind up close to the house. Teach someone a lesson."

"We have no proof for any of that, Lurleen," I said, "but you're right about one thing. Erin looks tall enough and fit enough to load a dead body into a kayak."

"A kayak she bought. And remember how upset she was about the blood—maybe she was worried it was her blood."

"You're giving me an idea, Lurleen. Let me call Mason."

I reached him on his cell. "Did you get any results from the sample, Mason?"

'Yes. It's human blood and we can probably type it."

"Any chance you can tell whose blood it is?"

"Nope. Even if we could, it would take months. The DNA folks are backed up with cases."

"Thanks." I filled Mason in on what we'd learned and then said goodbye.

"No chance of figuring out who the blood belonged to," I told Lurleen, "but I'm not sure how many people would know that. We might suggest we can identify the blood through a DNA match and see who crawls out of the woodwork."

Lurleen smiled and then she frowned. "That sounds dangerous."

"We won't do any of that until we talk it over with Mason and Danny," I said. "In the meantime, let's see if we can catch Chloe between customers, see if she'll tell us who the other woman is."

"My bet is on Lisa," Lurleen said.

"Or Shannon, the first wife," I said. "She's been around here as well."

"Yes, but she's hardly young."

"Neither is Lisa," I said.

We waited inside until Chloe finished with a customer. She looked around for someone else to assist and made no move to approach us. When we walked toward her she disappeared behind a case of jewelry as if she suddenly needed to organize it.

"It's our annual inventory, no time to talk," she said as Lurleen leaned over the counter and watched her work.

"Inventory in the middle of summer?" Lurleen asked. "In the middle of your busiest season? I don't think so. I'm an accountant, so I know that doesn't make sense."

Chloe stood and locked the cabinet door.

"We only need a few minutes of your time," I said." Maybe you heard your grandmother speaking to us."

"I heard a little," Chloe said, "but Grandma gets confused these days."

"Are you saying Erin didn't discover Brad with someone else?" Lurleen asked. "Mary seemed pretty certain of that fact, and Erin didn't deny it."

That was a lie on Lurleen's part—well, maybe not a technical lie. We never asked Erin that question because she claimed she and Brad had plans to marry.

Chloe looked surprised. "Erin told you about my seeing Brad with Lisa?"

Aha, Lurleen's fishing expedition was successful! Lurleen gave me a tiny smile.

"I tried to tell her it was probably nothing," Chloe said, "that they had to talk things over. They were getting a divorce, but when she pressed me, I told her I caught them in bed."

Chloe turned to us looking furious—almost as if she'd been the one who was betrayed.

"Brad was such a smooth talker and such a jerk! I told Erin she was lucky to be out of it. Then three days later, Brad winds up dead."

"Are you saying Erin could have killed him?"

"No! Of course not! Why would you even suggest such a thing? Brad hurt plenty of people—not just women. All his phony business deals cost people around here a lot of money."

"Can you give us some names?" I asked.

"Howie, for one—Brad's younger brother."

"I thought they weren't speaking to each other," I said. "And then suddenly he turns up and pays for a wake for his brother at a local bar. I couldn't understand any of that."

Chloe nodded. "Howie isn't much better than Brad. He's like a worm that surfaces after a storm. He knew about the divorce and came back to see what he could get out of it—maybe try to worm his way into Lisa's good graces."

"He didn't sound fond of Lisa when he talked to us," I said.

"Didn't Howie come on to Shannon before Brad's first wedding?" Lurleen asked. "I thought that's what caused the fracture between him and his brother."

"Yes, it did. My grandmother told me about it. But Howie was like Brad in one important way—no women were off limits for either one of them."

"I wonder if you know about other people who hated Brad," I asked, "people we don't know about yet?"

"I don't know who's on your list," Chloe said, "but Erin's father should be. Scott Huxley invested a lot of money in Brad's schemes."

"We heard that Brad tried to make things right with Scott by practically giving his boat to him in a very sweet deal," Lurleen said.

"I don't know about that," Chloe said, "but I do know he wasn't happy when he found out about Erin's relationship with Brad."

"So he did know about that?" I said. "How did he find out?"

"I might have said something," Chloe said. "I didn't want to see Erin throwing her life away on someone like Brad. All he did was hurt people."

"Hmm, you sound so angry," Lurleen said. "Your grandmother told us she invested in Brad's vision for Apalachicola—the one that would develop Apalachicola into a cultural center for artists."

"That's hard to believe," Chloe said, "but my grandmother always did have a mind of her own."

A customer walked in and Chloe looked relieved. "I have to get back to work."

We left the store.

"Chloe was certainly quick to defend her friend," Lurleen said, "a little too quick. I think the girl doth protest too much. I'm not sure who she's trying to defend."

"She sounded so furious at Brad," I said, "that struck me as odd."

"You think they're in it together?" Lurleen asked. "Two girls dumping Brad's body into a canoe would be a lot easier than one."

"I don't know." An idea occurred to me. "Chloe's a very pretty girl. Any chance Brad was ever involved with her?"

"Why would you ask that?" Lurleen said.

"I don't know. She just sounded so upset. Maybe it was because of what Brad did to Erin, but what if it was about what Brad did to her?"

"Like maybe she was involved with him romantically?" Lurleen asked.

"Just a thought," I said. "Brad got around, and I wonder if there was any woman he didn't sleep with."

"I'll bet he didn't sleep with Dawn, our bookstore owner," Lurleen said. "She seems too level headed and not exactly his type."

"You mean not young enough, not beautiful enough?"

"I think she's very attractive," Lurleen said, "but she doesn't play that up. Brad seemed to like his women young or flashy. Dawn is neither."

"So if she wasn't involved with him, she might be the perfect source to give us more information about what's really going on."

"Exactly," Lurleen said.

Mason called before we could make our way to the bookstore.

"Need you back here," he said. "How soon can you get here?"

34

I looked at Lurleen. "Mason wants us home. Dawn will have to wait."

Lurleen nodded.

"We'll be there in forty minutes or so. Sooner if I let Lurleen drive."

"Let Lurleen drive," Mason said.

"What's going on?" I asked.

"It's better if we talk when we're all together. This phone service is too spotty, and another storm is headed our way."

I looked at the blue sky overhead. "It's clear here. Are you speaking metaphorically or is there an actual storm on Cape Azul?"

"Storm clouds overhead look about to burst," Mason said, "literally and figuratively. Just come on home, Ditie—we'll do better if we're all in one place."

"We're on our way."

We ran to our car, and Lurleen got us home in twenty-five minutes. I clocked her. Fortunately, no police did the same.

We ran up the outdoor stairs and called out as we entered the condo. Danny and Mason were seated at the dining room table drinking coffee.

Mason looked at his watch. "That was fast. Thanks, Lurleen."

"*Mon plaisir, chéri.*"

"What's up?" I asked. "No rain that I can see."

"Scott fired Danny," Mason said. "He got word that we took a sample of blood to Tallahassee—maybe from one of the techs there who wondered why we were bringing in the sample. Scott was furious. He said we had no right to do that, and that he'd sue us for obstructing an investigation."

"Can he do that?" I asked.

"I doubt it," Mason said, "but he wants us gone from the Cape."

"We're not going are we, Danny?" Lurleen asked.

"No. I told him we'd be staying through our rental contract, and he said he'd see about that. The real question is what we do next."

"If he got that excited about the blood we found, it must be important," Lurleen said. "He must be protecting someone—himself or perhaps his daughter."

We heard a crack of thunder, and I stood up and walked to the sliding doors leading to the deck. Rain drenched everything within minutes. Mason came and stood beside me.

"What is it, Ditie?" he asked. "You have that plot-hatching look in your eye. What are you thinking?"

"I did have a thought," I said. "I know we can't match any person to the blood we found, but what if we say we can? What if we say you have a special contact in the DNA network that will let you do that, Mason? Maybe that would flush someone out."

"Or maybe it would put us all at risk," Mason said. "It's not our fight. I have no jurisdiction here. Scott Huxley is in charge, and he's made it clear he doesn't want us around."

Danny and Lurleen joined us to look out at the rain that was washing over our deck, saturating the towels and shoes we'd left outside.

"It might not be your fight, Mason, but it sure as hell feels like mine," Danny said. "I trusted the guy, and then when I get too close to the truth he fires me. He was playing me for a dupe the whole time. I don't like that. Maybe we should see who gets nervous when we say we're fast tracking the DNA evidence."

Mason ran a hand over his bald head. "It's not the way I do things," he said, "but it would let us see who's on the side of finding Brad's murderer and who's not."

He moved from the window to the kitchen as if to step away from the storm and clear his head. He poured himself another cup of coffee—perhaps to give himself a moment to think. "It could be dangerous, very dangerous. We came here for a celebration—not a murder investigation."

I stood beside him and put a hand on his arm. "I don't think you can let this drop."

"Could I persuade you and Lurleen to go home," he asked, "and let us handle it?"

I shook my head. "We all go or we all stay. I'm not losing you after a week of being married."

"We have the kids to think about," Mason said.

"I never stop thinking about the kids," I said. "If you think one of us might be harmed by the scheme I'm suggesting, then we don't do it. We go home instead."

Danny kept shaking his head. "I won't walk away and let some murderer run free. I've never done that."

Lurleen had stayed quiet, which was unusual for her. She'd remained at the window, staring out as the rain pelted down and then subsided.

"Look," she said. "a rainbow, a double rainbow, and Ditie, you won't believe this—I see dolphins."

I grabbed a set of binoculars and saw them too. The sun was out as if there had never been a storm.

"Let's leave this investigation alone for a while," I said, "and let's get in that water. Maybe swim with the dolphins."

I knew that wasn't going to happen—the dolphins were always farther away from shore than they looked—but I also knew we didn't need to make any quick decisions. Maybe Brad's killers were the strangers the bartender claimed he saw—if so they were long gone. If the murderer was closer to home, then we could take time to think about our next step. Scott had fired Danny, so he was off the case, and Mason had been warned not to interfere in any way. The murderer wouldn't consider us a risk any longer.

We could act like normal beach-loving adults for now. We changed into bathing suits and headed for the water. The sun was bright, but the storm had left the water colder than I liked it. The dolphins had disappeared by the time we got in the water. If anyone saw us, they would have assumed we were having the times of our lives, and for a little while we were.

After an hour we headed back inside to search for a quick and easy dinner. Danny ran out to buy fish and shrimp. He cooked up a batch of paella, and we savored every bite.

After dinner we watched an old DVD, and I found myself drifting off cuddled next to Mason.

"Bedtime," he said looking at me.

I woke up for a moment. "I think we have to mull over our plans," I said. "I make better decisions after a good night's sleep. Let's see what we all want to do in the morning."

As it turned out, we didn't have to make that decision. Storm clouds appeared once more and erupted during the night. The literal ones, and the ones that solved a murder case.

35

Before we went to bed, we called the kids and told them we'd be home in a few days, maybe sooner. They sounded a little disappointed. Melissa told us to take our time. She was having fun, and it seemed the kids were as well.

Mason and Danny made sure the house was locked up tight, curtains drawn, the wooden wedge put into the footer of each sliding glass window upstairs and down.

We didn't want any surprise visitors. Scott had been nervous enough about the investigation to fire Danny for whatever reason, and he'd threatened Mason with a law suit. Still, no one would be crazy enough to harm an Atlanta city detective.

Or would they?

In addition to everything else he did, Mason set up a barricade on the ground floor near the front door. No one would get in or out without our hearing their attempts.

I was sure we wouldn't sleep, but I was wrong about that. Mason and I snuggled in bed and talked about the kids. It became clearer and clearer that we couldn't put our lives at risk out of some desire for justice. Who knew where justice lay anyway? Brad had misused a lot of people. That wasn't to say he deserved to be murdered, but it was to say maybe this wasn't our fight.

"We're agreed then?" I asked Mason. "We go home tomorrow?"

"I think we do," Mason said. "I can't predict what Danny and Lurleen will decide to do. Danny's his own man and he's pretty ticked off."

"If Danny doesn't go," I said, "Lurleen won't leave."

"I know," Mason said. "Let's cross that bridge tomorrow. Listen to that rain. Only a fool would come out in that kind of downpour."

No one tried to break in that night. It was a gentle knock that woke me. My super power—or perhaps curse—was my very acute hearing. I looked at the clock—it was five in the morning. I woke Mason.

We didn't turn on a light, and Mason grabbed his gun from a drawer in the nightstand. We walked downstairs together, and at the foot of the stairs I heard him click off the safety. He motioned me back to the half bath where I could disappear into the darkness and lock a door if I needed to.

"Who is it?" Mason called standing to one side.

'It's me, Lenny. Let me in. Please!"

Mason flipped on the porch light, peered outside through a window to the side of the door.

"It's Lenny and Chloe," he told me. "I'm unlocking the door."

I nodded.

They stood on the porch, soaking wet, looking like rug rats.

"Come in," Mason said. He put his gun in his pocket.

I handed each of them a towel from the bathroom.

"I'm going to wake Danny and Lurleen," I said.

"Fine," Mason said. "I've got this."

Danny and Lurleen were dressed and downstairs in five minutes. Chloe was still shivering.

"I'm making coffee for everyone."

I could do that and still hear the conversation.

"Why are you here?" Mason asked.

"It was the only time Chloe could get out of her grandmother's apartment without Mary noticing. Mary's not a heavy sleeper until the early morning hours."

"You still didn't answer my question," Mason said.

"We had to come," Lenny said. "I was the one who brought over the kayak the other night because I needed you to find the blood inside, and you did. I saw Scott in that kayak the night before Brad's body washed ashore."

Mason waited.

"I didn't think anyone was with him," Lenny said, "I didn't see a second head."

"How far off shore was he?" Mason asked. "What time of night? How did you know it was him?"

"I knew it was him because he told me Erin had bought the boat, so the two of them could spend time together. I knew it wasn't Erin in the kayak. You couldn't miss her long blond hair even from a distance. It had to be Scott."

"How far out?" Mason asked again.

"Two hundred yards or so."

"So all you could see really was a head from that distance," Danny said.

"What are you saying? You think I came here at five in the morning to make up a story? Why would I do that? Why would I have brought the kayak to you in the first place?"

Mason ignored the questions. "What time was it when you saw the kayak?"

"Just after sunset. It was still light enough to see. First, I thought it was a crazy time to be out in a boat, then I recognized Scott. He has this brilliant yellow drysuit—his uniform when he's on the water. It picked up the last remnants of the sun—made him look like some kind of God. I know that's why he bought the suit in the first place."

"Let me get this straight," Danny said. "You saw someone in a kayak after sunset and you assumed it was Scott Huxley. How did you come by the kayak after Brad's body was found?"

"Chloe found it—under a tarp near Scott's house when she was visiting Erin."

"You found the blood in it?" Danny asked.

Chloe shook her head.

"I found that," Lenny said. "That's when I knew I had to bring you two into the investigation."

"You think Scott killed Brad—that's what you're saying," Danny said.

"I guess that is what I'm saying. He was always looking for strangers to blame Brad's death on. He never wanted to look closer to home. Then he fired you. At that point, I thought he couldn't be trusted."

"He hated Brad?" I asked.

"More like he loved Lisa and Erin," Chloe said. "I told Scott about Lisa sleeping with Brad, and he was angry. Angrier even than when he found out Erin was seeing Brad. When Brad turned up dead, I think Lenny and I both wondered if Scott had done it."

"Did you wonder if Erin might have been involved?" Lurleen asked. "She was the one most hurt by the discovery."

Chloe was silent for a moment. "Erin couldn't do a thing like that."

"Could you?" I asked. "You seemed so furious at Brad. It almost sounded as if he'd betrayed *you.*"

Chloe blanched. "We come here to tell you the truth and you make it sound like we're the ones who are guilty. I didn't kill Brad."

"Did you love him?" Lurleen asked. She asked it gently. "It seems he had an uncanny way with women."

Chloe didn't answer.

Lenny backed away from her. "Tell her you didn't! Go on, Chloe. Tell her you wouldn't be caught dead with Brad."

Nothing from Chloe.

"Let's take a minute," I said. "Sit at the table and I'll get the coffee."

I poured coffee for everyone.

Lenny was the first to speak. "Chloe, you didn't fall for that guy, did you? You couldn't have—you saw how he treated Erin."

"How he treated Erin?" Chloe said. "He treated her like a queen. He was going to marry her."

That left all of us speechless.

"What about seeing Lisa and Brad in bed together?" Lurleen asked.

Chloe hung her head. "I made that up. It wasn't true. I wanted to get Erin away from Brad. Erin was about to get everything I wanted."

"So you *were* involved with Brad?" I asked.

My relationship with Brad was over almost before it started," Chloe said. "I think he was using me to get to Erin."

"Did Erin know you loved Brad?" Lurleen asked.

Chloe shook her head. "I never told her, and she never knew."

"That must have been a hard secret to keep," Lurleen said.

"Yes, but I thought that what Brad did to me he'd eventually do to Erin."

"Then, you found out they were going to get married," I said.

Chloe looked devastated. "Yes. Erin told me Brad had changed, that he said he'd never really loved anyone before her." Chloe stopped for a minute and rubbed her eyes, pushed her wet hair off her face. "That was so hard to hear. Erin was going to get everything I wanted. I couldn't bear it."

"So you lied to Erin," I said.

"Yes."

"Did you and Erin have anything to do with Brad's death?" I asked.

My question was followed by a very long silence. "I didn't," Chloe said, "but I wonder if Erin might have. She told me she needed to talk to Brad. She was going to take the kayak out and confront him on his boat. When you told me you found blood in the kayak, that was when I began to wonder."

36

"Why did you come to see us at this hour?" Mason said.

"I thought we had to come," Lenny said. "I was afraid you might take off. I knew Scott had fired Danny and that he was threatening you with obstruction of an investigation. I knew it would take an outsider to solve this murder, and I didn't want the two of you to leave."

"A murder you think Scott or Erin committed?" I asked.

"Maybe," Lenny said. "Probably."

"Maybe they did it together," Chloe said.

"We just want to see justice done," Lenny said.

"Really?" Mason asked. "All you've done—all you've both done—is plant seeds of doubt about Scott and Erin. Was that your intention? You two had as much reason to kill Brad as anyone else."

"What the hell are you talking about?" Lenny asked. He looked angry—angry enough to kill, despite that trite expression. He must have seen me staring at him because he quickly settled down. "I knew nothing about Erin—or Chloe for that matter," Lenny said. "I never understood Brad's pull with women. He seemed to get the girls I wanted. And then he discarded them as soon as he got them. But I didn't kill him if that's what you're driving at."

Chloe looked at him. "Which of us did you want, Lenny—me or Erin?"

"I wanted you. How could you not know that? I asked you out a dozen times."

"Always as a friend, Lenny. Always, so we could talk about Erin."

"That's not true," Lenny said. "You acted like I wasn't good enough for you, like you were getting out of Apalachicola as soon as you could."

"Maybe I did want more, Lenny. I saw you as a small town cop with no ambition. I guess that was Brad's appeal. He promised me the world and then later he offered me an emerald and diamond bracelet."

"But he never actually gave you that bracelet," I said, "did he?"

Chloe shook her head. "He showed it to me; he promised it to me and then he changed his mind."

"So who was going to get that bracelet?" Lurleen asked. "We found it in a box wrapped like a present."

"I assumed he meant to give it to Erin," Chloe said.

"We heard about another emerald and diamond bracelet," Lurleen said, "one your grandmother made years ago. You sure he meant to give you the big one?"

"I saw it—it was the one you showed me in the picture. I knew what it meant when he offered a present like that—that he was done with me," Erin said. "Then he took it back, and I hoped he might give it to Erin to say he was done with her."

"Maybe," Lurleen said, "but why would he have shown it to both of you and then tucked it away?"

"Brad always liked to pretend he was generous," Chloe said. "Maybe he meant to get rid of both of us with one bracelet."

"Or maybe he changed his mind about what he wanted to

do with that bracelet," I said. "Maybe this time his gift was meant to be a sign of his lasting love—that's what Erin thought. Then he was killed before he could give it to her."

"That bracelet could have interested a lot of people," Mason said. "It was likely worth a fortune."

"For some," I said, "it would have meant a great deal of money, and for others, the ultimate betrayal."

"People like Lisa for example," Danny said. "I never believed a word Lisa said except when she talked about Brad—then I did believe her. She wanted him back. Some of it was for the money, but some of it sounded like she was still in love with him. So, when she heard he intended to give the bracelet to Erin at their wedding, she would have been furious."

"You're suggesting Lisa killed Brad, stuffed him in the kayak, dumped him at sea?" Mason asked. "She couldn't have done that on her own."

"Maybe Scott helped her," Lurleen said. "He owned the kayak. He might have hoped once Brad was gone, Lisa would turn to him. Kill two birds with one stone—get Brad permanently away from his daughter *and* help Lisa dispose of the body, if she was the one who killed Brad."

"What about the investigation Lisa pressed Scott to do?" I asked. "She acted as if she really wanted to know who killed Brad. Do you think that was a scam that both of them were in on?"

"I think we need to talk to Lisa," Mason said. He looked at his watch. "It's five-thirty—maybe we track her down when it starts to get light. She said she was staying on the Cape. Do either of you know where she's staying?"

Lenny gave them the address. "I don't know if she'll be home, and I don't know if she'll be alone."

"You mean you think Scott will be there with her," Mason said.

"Wouldn't be surprised."

"Anything else you need to tell us before you leave?" Mason asked.

Lenny shook his head. "We want justice, that's all."

"Why would you care so much about that?" I asked.

"We don't want a murderer to go free," Chloe said, "like we said, and it was clear Scott wasn't going to look into things anymore."

"It's mostly out of our hands," Mason said. He paused, looked at me, and then said he knew someone in the Tallahassee forensics lab who was fast tracking the DNA on the blood sample we left. "They'll get the answer even if we're not in town to hear about it."

Did Lenny blanch? I couldn't tell for sure. Chloe seemed to notice the look on Lenny's face.

She grabbed her rain coat as if she were eager to get out of our place. "Never mind then," she said. "If you want to leave, you should go."

"What?" Lurleen asked. "You come here at five in the morning to beg us to stay, and now you're telling us to leave. Why the change of heart?"

Chloe looked at Lenny, and Lenny jumped in. "If the police are looking into the DNA, perhaps the case can be solved without your help. You can leave. If anyone else hears about this, I mean the real murderer, they might get desperate —you could be in danger."

We watched them drive off.

"Did we learn anything useful?" I asked.

Lurleen shook her head. "They want justice at five o'clock in the morning? Then, half an hour later, they want us to leave. More likely they wanted to know what we knew, what we were up to, what our plans were—that's what I think."

"You might be right," Mason said. "They ended up sounding more like suspects than people in pursuit of justice —justice for a man they both hated."

"I wish I could be in that car to hear what they're talking

about," I said. "You think one of them will tell Lisa we're coming to her house?"

"Maybe," Mason said. "Maybe we ought to try to get there first."

None of us was going to sleep anyway, so it didn't sound like an unreasonable request.

"Let's see who she's with if she isn't alone," Lurleen said. "I've got her address."

We piled into the car and headed north toward the entrance to the park.

I looked at the address. "Did you realize the address is inside the park? The park will be closed at this point—we won't be able to get in."

"Then we'll go on foot," Danny said. "We can park near the entrance and walk on the beach. This address is for one of the first big houses inside the park. The storm's over, and going on foot will give us a chance to surprise her."

37

We drove to the entrance to the park at the north end of Cape Azul and saw the barrier and sign indicating the park was closed. We parked off the road and walked down a path to the beach. It was still dark and we picked our way along the beach with flashlights, on the lookout for any turtle footprints in the sand. We didn't spot the heavy body markings of pregnant mothers or the scurry of little baby turtles into the sea. It was just the four of us and lots of untouched sand. We didn't speak. Danny pointed out the house to us. It was large on steel pilings and most of the house was out of view beyond the dunes that lay in front of it.

Mason motioned us to follow him up the wooden walkway to the house, and we did. Black-out curtains were drawn, but it looked as if there was light we could see from a side window. We crawled below the side window. Lurleen poked her head up as she passed the window and then motioned us to move beyond the front door.

Near the street, she whispered to us. "You know who's in there? Lisa and Mary."

"Mary's there?" That shocked me.

"Mary's there, and that's not all. Chloe and Lenny are

inside as well. They must have come straight here to warn Lisa."

Danny pulled out his gun, but Mason motioned for him to put it away.

"You go up those steps with a gun, and anything could happen. Put it away. Ditie, call Lisa. Tell her you have some news about Brad's death that can't wait. Let's see what happens next."

I called. Lisa pretended she'd been asleep. "This can't wait until morning?"

"I know it's early," I said, "but I have news about Brad's murder."

"News about Brad?" she asked. "You know who killed him?"

"It's new evidence, and I thought you might be able to help me put the pieces together."

"You're coming here alone?" Lisa asked.

"Lurleen is with me. Danny and Mason had to go to Tallahassee."

"All right. Let me give you my address."

Thank goodness she said that—I might have forgotten to ask. "We'll be there in fifteen minutes."

She gave us the code to raise the barrier and enter the park.

We waited outside in the dark away from the street lamp. We expected to see a mass exit, but that didn't happen. After ten minutes, Lurleen and I walked up the steps to the house.

We knocked and Lisa opened the door. "I didn't hear your car," she said. She glanced at the driveway. "Where is it?"

"We didn't want anyone to see it in case someone was watching your house," I said.

"You think someone is watching me? You think my life is in danger?"

"We aren't sure," Lurleen said. "May we come in?"

Lisa ushered us inside. We'd had time to plan our strategy.

"No one is here with you?" Lurleen asked.

"No, why would you ask that?"

"I apologize," Lurleen said, "but we heard you and Scott were spending time together. You know how rumors start."

She laughed at that. "We're not seeing each other. Now, tell me what was so urgent you had to see me at this time in the morning." She looked at her watch. "It's barely six a.m."

I told her the truth—to a point. "Chloe and Lenny came to see us an hour ago, and their story didn't make a lot of sense. I thought you should know."

"They told me Danny and Mason were at your house," she said.

"So you know they came to us," I said.

"Of course I know. We're close."

"Did they tell you they suspected you might have had something to do with Brad's death?" Lurleen asked the question as if she were full of concern for Lisa.

"That's absurd."

I wasn't sure she thought it was absurd by the look on her face.

"How did you know about their visit to us?" Lurleen asked. "Did they stop by here?"

"They called."

"That's odd," I said. "The phone service where we're staying is very spotty. I was lucky to reach you, and I couldn't get through to the kids last evening."

Lisa motioned to the couch, but Lurleen and I remained standing. She did the same.

"You must be so anxious to get home to them," Lisa said. "I assume you'll be leaving soon?"

"We'd like to finish things up here first if we can," I said.

"What are you talking about? What is it you want to tell me about—or ask me? If you think you know who killed Brad, that could be dangerous for you. You should speak directly with Scott. You come here when it's still pitch black outside.

That's a crazy thing to do if there's a murderer on the loose—why are you here?"

"As we told you," Lurleen said, "Chloe and Lenny came to us an hour ago, and their visit didn't make a lot of sense. They implicated you—and we thought you should know that."

"Enough!" Lisa said. "Chloe. Lenny, get out here."

The two of them must have been standing in the long corridor that led away from the great room where we were. They came out to stand beside Lisa, heads down.

"Let's get this cleared up," Lisa said. "You said something to Lurleen and Ditie that made them think I could have been involved with Brad's death. I loved Brad, even at the end, so you know that's impossible."

Lenny and Chloe didn't seem to know which way to look.

"Aren't you leaving someone out?" Lurleen asked. "I saw Mary here when I looked in your window."

"You looked in my window?" Lisa said. "You've been watching me? I'm calling Scott. You could be arrested for prowling around this place."

"Please call," I said.

Lisa sat down on the couch instead. "We can sort through this by ourselves—I know we can."

Mary silently appeared from the hallway. She didn't greet us.

"Is that it?" Lurleen asked. "Anyone else lurking in the wings?"

"No," Lisa said. "We've been meeting here almost every night since Brad died. At first it was because we wanted to find out what had happened to him, and we didn't think Scott was doing an honest investigation. But then that changed. We began to talk more about Brad—about the damage he'd done to each of us."

Mary shifted in her seat. "We began to admit that whoever killed him did us a favor."

"Did you hate him?" Lurleen asked Lisa. "You said you wanted him back, didn't want the divorce after all."

"That was before—"

"Before what?" I asked.

"Before I found out about Erin—that he was intent on marrying her and needed to get divorced to do it."

"Everyone in this room had a reason to kill Brad," I said. "Did you do it together? Am I revisiting the Orient Express, only this murder didn't require a moving train?"

That appeared to shock all of them, including Lurleen.

She looked at me. "Am I sitting in a room full of murderers?" she asked.

"Don't be absurd," Mary said. "Do I look like a murderer? Do I have the hands of a murderer?" She showed us her swollen joints. "I can barely do my jewelry anymore. I couldn't do any harm to Brad."

"Unless you were the mastermind," Lurleen said.

"It's likely it took more than one person to kill Brad," I said, "or at least to dispose of his body."

Everyone remained silent.

Lenny finally stood. "You come here, the two of you, and accuse us of murder. Why wouldn't we kill you and take you out to sea if we were the murderers?"

Lenny pulled out a gun. Government issue.

That made everyone in the room freeze.

He waved it around before he laid it on the table."My point is that you barge in here, make your accusations. You know nothing about us, and you think we're murderers. We're not. We came to see you to find out what your plans were. We tried to make it seem as if we wanted you to stay, but we hoped we could make you nervous enough to leave. I admit our plan backfired. None of us is guilty of murdering Brad—but that doesn't mean the thought never crossed our minds. "

"Okay, let's say none of you killed Brad," I said. "You're telling me now you don't care who did."

"Whoever killed Brad had a good reason to do it—maybe a very good reason." This was from Mary.

"May I ask you something I'm curious about before we leave?" I looked at the four of them.

"Go ahead," Lisa said.

"The two bracelets—the two emerald and diamond bracelets. Why two? One made by you, Mary, years ago, and the other made somewhere else more recently. We know Brad liked to give a piece of jewelry to the women he was leaving—to soften the landing perhaps. We know where one bracelet is, but where is the other?"

The women shook their heads.

"If we knew who got that other bracelet," I said, "maybe that might lead us to our murderer."

Lisa spoke first. "I think the wrapped bracelet was meant for me, and Brad died before he could give it to me. I think it was his way of saying good bye to me—forever."

"Perhaps," I said. "Unless, it was meant as a wedding gift to Erin. Either way, where is the other one?"

Chloe's face went red and she jumped up. "He didn't give the other one to me. I told you he showed the larger one to me and then took it back. He said it belonged to someone more deserving. Maybe it was for Erin, that's what she thought."

There was only one other woman close to Brad that we knew about. Shannon. She'd thrown the old bracelet in Brad's face after the first divorce and she'd turned up for the second divorce proceedings determined to get her fair share—at least that's what people told us.

The front door opened, and we all jumped. Danny and Mason entered the house, guns drawn. Between them were Shannon and Howie.

"We found these two snooping around your house," Mason said.

"We thought you must have the bracelet, Lisa," Shannon said. "We came to get it back—it was supposed to be mine.

Then before we could ring your bell, we ran into these two." She pointed to Mason and Danny. "We wanted to catch you alone. We knew you hated Brad as much as we did. Brad was an equal opportunity offender. He was happy to take advantage of women *and* men with his charm and his phony promises."

Lisa looked outraged. "You expect me to believe you were going to ask nicely for the bracelet, and I'd give it to you. Scott Huxley has the bracelet—somehow you two didn't know that?"

"Scott stayed very closed mouth about the investigation," Howie said. "It made me wonder what he was up to."

"Brad was a cheat to the end," Shannon said. "He tried to give me that old scrawny bracelet once more in order to get me off his back and out of his life. But I knew about the big one. I deserved the big one."

"Brad might still be alive," Howie said, "if he'd parted with that one."

"Shut up, Howie," Shannon said. 'You've always had a mouth that was bigger than your brain."

He ignored her.

"You figured it out, didn't you?" Howie said to Mason. "You, the fancy detective from Atlanta. We only wanted what was ours. We never wanted to kill anyone."

"Stop talking, Howie," Shannon said. "They don't know a thing, They can't prove anything."

"I wouldn't be sure of that," I said. "The blood we found in kayak has been sent off for DNA identification. I'm pretty sure you're in the system, Howie."

Neither one of them knew enough to realize that wasn't the truth.

"If you want to tell us what happened," Mason said, "it will go a lot better for you. I'll give Scott a call, so he can hear it too."

38

"I'm calling my lawyer," Shannon said. "Not another word, Howie, or I swear I'll kill you myself."

Scott arrived ten minutes later with his deputy.

"Do you want to talk here or at the office?," he asked.

"I called my lawyer," Shannon said, "but it seems he's on vacation."

"So?" Scott asked.

"We can talk here," Shannon said, "but I won't have much to say. Not until I speak to my lawyer."

"I see," Scott said. "At least we can get started." He read them their rights. "Coffee, Lisa?"

Apparently, in a small community the chief of police could do what he wanted.

"They know what we did," Howie said. "I'm not waiting on a lawyer."

He turned to Scott.

"We didn't mean to kill Brad. I didn't anyway—he was my brother for god's sake. Shannon got me into this."

"Shut up, Howie," Shannon said. "You were a fool when I first met you, and you're still a fool."

"I won't shut up. I wasn't the one who bashed him over

the head. I was just the one who dumped his body—where she told me to."

"None of that is true," Shannon said. "Brad and I were getting back together. Howie was afraid he'd be cut off and left with nothing. He killed Brad."

That's how the first twenty minutes of Scott's interview went. Back and forth, but through the cracks the truth emerged.

Shannon came back to town when she heard about the second divorce, not out of love but out of greed. She wanted her share. And there was money to share. Howie provided that source of information.

Brad had finally found a way to get out of debt—he did drug running for some drug dealers. Those were the men the bartender had seen Brad talking to. They were the same men who came to the impromptu wake for Brad.

Brad had called his younger brother back to town to offer him a deal. He actually did believe Erin was the love of his life and he wanted to clean up his act. He offered Howie the business as long as he could keep the money he'd already earned. Howie was happy to agree to that and to take over the job. The only problem was that Brad had already given the boat to Scott as proof of his love for Erin, and he hadn't bothered to tell Howie that part of the story.

"Sure I was mad about that," Howie said, "But it wasn't going to be that hard to buy another boat. I had connections, other people who might want to be involved with some easy money. I knew I could have made a killing," Howie said with no sense of irony.

Hearing this was the breaking point for Shannon.

"You told me nothing about this! I wondered why you wanted to buy the old mansion off me. Then I thought maybe you actually believed in some of Brad's schemes about Apalachicola."

"I only believed in one of his schemes," Howie said. "The

one about drug dealing. That old house was going to make a great staging area for some serious drug distribution," Howie said. "Who would believe anything was going on there if I was careful. I could claim I was fixing it up if anyone got curious."

"And you thought you could leave me out of it!" she screamed. "I should have killed you, not Brad. You're the one who deceived me and thought you could get away with it."

She turned to the rest of us and kept talking.

"All I knew was that Brad was trying to get rid of me. The bracelet he showed me at first was magnificent. I could have settled for that and left him alone. Then the bracelet he actually gave me was the old piece of jewelry—jewelry made by Mary, no less. I never liked her stuff. What a fool he was.

"He thought he could con me into thinking it was the same bracelet. He didn't think I'd remember he'd tried to give it to me when he first took up with Lisa—I didn't want it then and I certainly didn't want it now. He tried to tell me how much I could get for the tiny diamonds and smattering of emeralds. I'd had enough of his games. We had a big argument about the way he thought he could con me. He laughed about it, told me to get a life, stop mooching off him and everyone else.

"When he turned away, I smashed him over the head with the closest thing I could find—the bottle of cheap scotch he'd offered me. Fitting. I was sick of him treating me like a second rate human being. He thought I was too stupid to tell one bracelet from another—a second rate wife easily disposed of.

"It all happened in the old Arborough house. I called Howie to help me. The place was falling down so we didn't have to clean up anything. Brad liked to meet there because no one would spy on him."

Shannon was done. She didn't seem especially upset about her confession. "It wasn't premeditated," she said. "So I spend a few years in jail. I've watched 'Orange is the new Black.' I'll be tough enough to survive." She looked around the room. "I did you all a favor. You should treat me like a bounty hunter—hire me a good lawyer at least. One who doesn't go on vacation when he's needed."

Howie looked plenty upset.

"I only did what Shannon told me to do," he said. "We wrapped Brad in an old carpet—just like in the movies. There were some tacks in the carpet, and I cut my finger—so when you talked about blood and DNA testing, sure I got nervous it might have been mine. We hauled Brad back to Cape Azul. I knew about the kayak—I advised Erin about the best one to buy—and I knew where she kept it.

"I found Scott's dry suit hanging beside the kayak. It was a no-brainer. If anyone saw me in the water, they'd assume it was Scott. He was off somewhere with Lisa, and Erin was most likely with Chloe. I dropped Brad at sea. I knew Shannon wanted Brad's body to turn up near the houses he owned with Lisa—she wanted to scare her."

Shannon sighed. "Oh, Howie. you always were the stupid brother. For the record, I had no intention of killing Brad. All I wanted was the bracelet, and he wouldn't give it to me. It was self-defense."

"Self-defense?" I said. "You bashed him in the back of the head."

"I knew he was going to grab a poker from the fireplace and attack me. A decent lawyer will prove that."

She sighed again. "I deserved that damn bracelet for the way Brad ruined my life. Howie and I searched for it in the big house and tried to make it look like some random break-in. Then we searched Scott's house when it was empty."

Scott went over the details slowly and carefully. He'd been on Brad's boat with Lisa at the time Shannon and Howie were at the big house. Erin had gone to stay with Chloe to recover from Brad's death, so the coast was clear.

Brad had given Scott the boat as proof of his affection for

Erin. He hadn't told Scott about his business on the side, only that he no longer wanted the boat.

The sun was coming over the dunes when Scott and his deputy left with Shannon and Howie.

We left shortly after Scott did.

Lisa smiled broadly as she showed us out the door. "If you want to rent one of my houses next summer, I'm sure that can be arranged."

"Thanks," I said, "but we'll stick with our old rental down the beach."

That's how the investigation ended. Plenty of people were happy to see Brad dead, but only one of them appeared fed up enough with his con games to commit murder.

39

Danny, Lurleen, Mason and I went back to our condo and sat on our deck as the day broke bright in a cloudless sky. We didn't say much.

"We might as well enjoy the two more days we have," Mason suggested.

"But I miss the kids," I said.

"Compromise?" Mason asked. "Isn't that what marriage is all about? Let's split the difference—enjoy today and head home tomorrow. Deal?"

"Deal," I said.

"What about you two?" I asked Lurleen.

She looked at Danny. "We haven't had a lot of time together, so we might stay until the end of the rental agreement."

Danny nodded. "Good plan," he said. "I'm for that."

"We'll pack up everything, Ditie," Lurleen said. "You and Mason can leave whenever you want tomorrow. Today, I need a little more beauty rest."

She and Danny headed upstairs. Mason and I followed.

"What a honeymoon," Mason said.

"What a wedding," I said back to him. "Do you have regrets?"

"About the wedding—absolutely not. About the honeymoon, only a little. You and I did get our time together. We are all safe, but maybe next year, we should take our anniversary in some more exotic spot."

"Somewhere like the South of France?" I asked. "Lurleen has promised we could spend time in her home in Aix-en-Provence. That should be safe enough since neither you nor I speak the language. What kind of trouble could we possibly get in?"

"You say that like a challenge," Mason said. "I'm never happy when you say it like that."

We spent the rest of the morning in bed—sleeping and doing other things. In the afternoon, we had a long swim together. We didn't see Lurleen or Danny until close to dinner.

Danny served us a feast of scallops in a rich cream sauce.

I woke up early the next morning and started to pack.

"Looks as if we're getting an early start back to Atlanta," Mason said when he finally opened his eyes.

"Do you mind?"

"Nope, I miss the kids as much as you do.'

We squeezed in one more walk on the beach and found a pod of dolphins in the water who were quite happy to put on a show for us. We were near the big house, and it was the perfect way to end our honeymoon.

I let Mason drive us home. I told Melissa we were headed to her house, and we all agreed we'd surprise the kids. She made sure they were home when we arrived at four.

I couldn't have asked for a better welcome.

"You're home, Mommy," Jason squealed. He sounded like a little kid at that moment, and he let me hug him for as long as I wanted to.

"Mama," Lucie said, hugging me tight. "I knew you'd be safe, but I'm really happy to see you."

Mason got his share of hugs. We said goodbye to Melissa and Noah and dropped Hannah at her house. We were home at our place by five thirty. If I thought I'd gotten a good welcome from the kids, that was nothing compared to Hermione's response. She was beside herself, licking all of us, jumping up and down, waggling her entire body. We took her outside until she could calm down. Majestic waited demurely inside the screen porch. When I entered he approached me and rubbed against my legs. When I sat down he hopped onto my lap and purred.

Our pet sitter told us how good both the animals had been. We thanked her, paid her and watched her walk across our yard to her own house.

"Home at last," I said. "I love being away, but being home with my family is where I really want to be."

Mason smiled. "Me too."

I'm sure we made quite a picture for anyone who might have looked in our bay window at all of us cuddled together on the sofa in our family room. We were a jumble of kids and animals. Mason and I were buried under the love. What a homecoming!

Lurleen checked in with us that evening.

"All settled?" she asked. "Danny and I will be heading home tomorrow. I got some great news today. You remember how I talked about starting a mystery bookclub? A friend emailed me to say she thought it was a wonderful idea. We could use her house, and she knew the people to invite. How does that sound, Ditie?"

"Sounds great to me," I said. "Reading mysteries might be a much safer way to enjoy a murder investigation."

"Yes," Lurleen said. "However, there might be a little more to it than that. You see one of the members hopes we can solve a mystery she seems to be very worried about."

"Oh, no, Lurleen," I said.

"Don't worry, Ditie," Lurleen said. "No one has died. Yet."

THE END

DITIE'S NEW RECIPES

DITIE'S WEDDING CAKE

Here's the recipe I used for Ditie's wedding cake. You could actually make this cake for fun and not for a wedding ceremony. Don't be daunted by the directions. (In other words, ignore all the design elements and focus on the ingredients themselves). I settled on three small cakes and a few cupcakes to share. I had plenty of delicious batter following directions for the nine inch cake. I cut the frosting recipe in half and had just enough to frost three small cakes and a half dozen cupcakes.

Simple Homemade Wedding Cake Recipe

★★★★★

4.9 from 34 reviews

• Author: Sally Prep Time: 5 hours (includes baking, cooling, decorating) Cook Time: 45 minutes Total Time: 6 hours Yield: 30-35 servings

This is a recipe for a completely homemade 2 tier wedding cake. For best taste and texture, and to avoid overwhelming your mixer, make each tier (cake) and each batch of frosting separately.

Ingredients

9-INCH BOTTOM TIER

• 1 and 1/2 cups (3 sticks, 345 g) unsalted butter, softened to room temperature

• 2 cups (400g) granulated sugar

• 3 large eggs + 2 additional egg whites, at room temperature

• 1 Tablespoon pure vanilla extract (yes, Tbsp!)

• 3/4 cup (180g) full-fat sour cream, at room temperature

• 3/4 cup (180ml) whole milk, at room temperature

6-INCH TOP TIER

• 1 and 3/4 cups (207g) cake flour spoon and leveled

• 1/4 teaspoon salt

• 3/4 teaspoon baking powder

• 1/4 teaspoon baking soda

• 1/2 cup (1 stick; 115g) unsalted butter, softened to room temperature

• 1 cup (200g) granulated sugar

• 1 large egg + 1 additional egg white, at room temperature

• 2 teaspoons pure vanilla extract

• 1/2 cup (120g) full-fat sour cream, at room temperature

• 1/2 cup (120ml) whole milk, at room temperature

FROSTING FOR 9 INCH CAKE & PIPING

• 2 cups (4 sticks; 460g) unsalted butter, softened to room temperature

• 8 cups (960g) confectioners' sugar

• 6 Tablespoons (90ml) whole milk or heavy cream

• 1 and 1/2 teaspoons pure vanilla extract

• 1/8 teaspoon salt

FROSTING FOR 6 INCH CAKE

• 1 cup (2 sticks; 230g) unsalted butter, softened to room temperature

• 4 – 5 cups (480-600g) confectioners' sugar

• 1/4 cup (60ml) heavy cream or whole milk

• 1 teaspoon pure vanilla extract

- salt, to taste

Instructions

1 Before beginning this recipe, watch the video tutorial above, read the entire blog post, read the recipe instructions, and review the recipe notes. Make sure you're prepared with the recommended special tools, which are listed right above this recipe .

(Remember: special tools are not necessary unless you are seriously making the cake for a wedding. If so, find this recipe online or watch the video: https://www.youtube.com/watch?....)

2 Make each cake separately. Start with the bottom tier (9-inch cake). Preheat oven to 350°F (177°C). Grease three 9×2 inch cake pans, line with parchment paper, then grease the parchment paper. Parchment paper helps the cakes seamlessly release from the pans.

3 Whisk the cake flour, salt, baking powder, and baking soda together. Set aside. Using a handheld or stand mixer fitted with a paddle or whisk attachment, beat the butter and sugar together on high speed until smooth and creamy, about 3 minutes. Scrape down the sides and up the bottom of the bowl with a rubber spatula as needed. Beat in the 3 eggs, 2 egg whites, and vanilla extract on high speed until combined, about 2 minutes. (Mixture will look curdled as a result of the egg liquid and solid butter combining.) Beat in the sour cream. Scrape down the sides and up the bottom of the bowl as needed. With the mixer on low speed, add the dry ingredients until just incorporated. With the mixer still running on low, slowly pour in the milk until combined. Do not over-mix. Whisk it all by hand a few times to make sure there are no lumps at the bottom of the bowl. The batter will be slightly thick. You'll have 8-9 cups of cake batter. Pour batter evenly into cake pans. Weigh them to ensure accuracy, if desired. Bake for around 23-26 minutes or until the cakes are baked

through. To test for doneness, insert a toothpick into the center of the cake. If it comes out clean, it's done. Allow cakes to cool completely in the pans set on a wire rack. The cakes must be completely cool before frosting and assembling.

4 Keep the oven on.

5 For the top tier (6-inch cake) prepare three 6×2 inch cake pans just as you did with the 9-inch cake pans in step 2. (Greasing with parchment paper rounds.)

6 Prepare the 6-inch cake batter the same exact way as the bottom tier. You'll have about 4 cups of cake batter. The batter will be thinner than the bottom tier's cake batter. Pour batter evenly into cake pans. Weigh them to ensure accuracy, if desired. Bake for around 18-22 minutes or until the cakes are baked through. To test for doneness, insert a toothpick into the center of the cake. If it comes out clean, it's done. Allow cakes to cool completely in the pans set on a wire rack. The cakes must be completely cool before frosting and assembling.

7 Using a large serrated knife, slice a thin layer off the tops of all the cakes to create a flat surface. This is called leveling off the cakes. Discard the leveled off piece (or crumble over ice cream!).

8 Make the frosting for the 9 inch cake (which is enough for the piping detail, too)**:** In a large bowl using a hand-held mixer or stand mixer fitted with a paddle attachment, beat the butter on medium speed until creamy, about 2 minutes. Add confectioners' sugar, milk, vanilla extract, and salt with the mixer running on low. Increase to high speed and beat for 2 minutes. Add more confectioners' sugar if frosting is too thin, more milk if frosting is too thick, or an extra pinch of salt if frosting is too sweet.

9 You can prepare the rest of the frosting now (step 11), if desired, or begin decorating the 9-inch tier. Place your 9-10 inch cake board on your cake turner. Place 1 cake layer on your cake board. Using a large icing spatula, evenly cover the

top with a scant 1 and 1/2 cups of frosting. Top with 2nd cake layer and evenly cover the top with a scant 1 and 1/2 cups of frosting. Top with the third cake layer. Using about 1-1 and 1/4 cups of frosting, spread it all over the top and sides in a thin layer, also called a crumb coat. A large icing spatula and bench scraper are helpful for applying the crumb coat. Set leftover frosting aside at room temperature. If you are decorating this cake in stages over 1-2 days, cover and refrigerate this leftover frosting.

10 Using the cake board to pick up the cake (be careful, it's heavy), place the cake in the refrigerator to help set the crumb coat as you work on the top tier.

11 Make the frosting for the 6 inch cake, just as you prepared the 1st batch of frosting. (Step 8.) Before adding any salt, taste the frosting, then add a pinch if desired. I always add a small pinch.

12 Place your 6-inch cake board on your cake turner. Place 1 6-inch cake layer on your cake board. Using a small icing spatula, evenly cover the top with a scant 3/4 cup of frosting. Top with 2nd cake layer and evenly cover the top with a scant 3/4 cup of frosting. Top with the third cake layer. Using about 3/4 cup of frosting, spread it all over the top and sides in a thin layer, also called a crumb coat. A small icing spatula and bench scraper are helpful for applying the crumb coat. Set leftover frosting aside at room temperature. If you are decorating this cake in stages over 1-2 days, cover and refrigerate this leftover frosting.

13 Using the cake board to pick up the 6-inch cake, carefully place the cake in the refrigerator to help set the crumb coat as you work on the exterior frosting on the top tier.

14 Remove the bottom tier from the refrigerator (along with the cake board) and place on the cake turner. Apply the remaining frosting for the bottom tier on the cake, reserving about 6-8 Tablespoons for piping detail. I recommend a large icing spatula and bench scraper to apply the outer layer of

frosting. For the decorative textured frosting as pictured, see my video above. As shown in the video, use a small icing spatula and run it around the cake while spinning with the cake turner. You will love this look because it's very easy, but resembles beautiful ruffles. Carefully lift the cake (with the cake board) and place on your serving platter or cake stand.

15 Wash and dry 4 cake dowels. Measure the height of the bottom cake. Measure and cut the cake dowels to match. Staying within a 6 inch circle diameter (you can use another 6 inch cake board to "imprint" a 6-inch circle in the very center of the top of the 9-inch cake), insert the dowels into the cake, spacing them about 2 inches apart to form 4 corners of a square. Push the dowels straight down until each touches the bottom cake board. Set aside.

16 Remove the top tier from the refrigerator (along with the cake board) and place on the cake turner. Decorate just as you did the bottom tier, including the ruffled textured frosting. I also turn the cake on the cake turner and apply this ruffled look on top of the 6-inch cake, too. This is optional. Along with the cake board, carefully lift the small cake and gently place it in the very center of the bottom tier.

17 Piping: After placing the small tier on top, you can pipe leftover frosting around the bottom of it. This covers up any spaces or smears that may have appeared when arranging the top tier. I pipe dots of frosting with Wilton piping tip #12 between the tiers as well as around the bottom of the whole cake. They resemble pretty pearls. To smooth down pointy peaks on these "pearls," moisten your fingertip with water and gently press down on the peak.

18 Add any decorative pieces, such as artificial flowers.

19 **Cake is best enjoyed within 3 days.** Refrigerate cake until ready to serve. After a few hours, the frosting will "set" and you can lightly cover with plastic wrap.

20 Cover any leftover cake tightly and refrigerate for up to 1 week.

Notes

1 Make Ahead & Freezing Instructions: Assembled cake is best enjoyed within 3 days. Refrigerate assembled cake until ready to serve. After a few hours, the frosting will "set" and you can lightly cover with plastic wrap until ready to display and serve. You can make the cake ahead of time before assembling it, too. The cake layers can be baked, cooled, and covered tightly at room temperature overnight. Likewise, the frosting can be prepared, then covered and refrigerated overnight. Let the frosting sit at room temperature to slightly soften for 20 minutes before assembling and frosting. (You may need to beat the refrigerated frosting in the mixer for 1-2 minutes if it's particularly solid. Feel free to add a Tablespoon of milk or heavy cream to soften it up as it mixes, too.) Crumb-coated cakes can be frozen up to 2-3 months. Cover with 2 layers of plastic wrap and 1 layer of aluminum foil. Freeze. When ready, thaw overnight in the refrigerator. Unfrosted cake layers can also be frozen up to 2-3 months. Bring to room temperature before decorating/serving. See how to freeze cakes for detailed instructions on freezing cake layers.

2 Why is everything at room temperature? All refrigerated items should be at room temperature so the batters mix together easily and evenly.

RUGELACH

This is a lovely recipe by Ina Garten that I found on the internet. It takes a couple of attempts to get it just the way you want it to be, but it's worth the effort. This is delicious and unlike any other pastry/cookie you'll ever make.

Keep the dough sticky (i.e., don't add too much flour) and you'll have mouth-watering results.

Look this recipe up on line and you can watch a video to see how it's made.

Enjoy.

Rugelach

- Level: Easy (I would probably call the level moderate. It takes a little practice to get it just right.)
- Total: 1 hr 55 min
- Prep: 10 min
- Inactive: 1 hr 30 min
- Cook: 15 min
- Yield: 4 dozen cookies

Ingredients:

8 ounces cream cheese, at room temperature

1/2-pound unsalted butter, at room temperature

1/4 cup granulated sugar plus 9 tablespoons
1/4 teaspoon kosher salt
1 teaspoon pure vanilla extract
2 cups all-purpose flour
1/4 cup light brown sugar, packed
1 1/2 teaspoons ground cinnamon
3/4 cup raisins
1 cup walnuts, finely chopped
1/2 cup apricot preserves, pureed in a food processor
1 egg beaten with 1 tablespoon milk, for egg wash

Directions:

1 Cream the cheese and butter in the bowl of an electric mixer fitted with the paddle attachment until light. Add 1/4 cup granulated sugar, the salt, and vanilla. With the mixer on low speed, add the flour and mix until just combined. Dump the dough out onto a well-floured board and roll it into a ball. Cut the ball in quarters, wrap each piece in plastic, and refrigerate for 1 hour.

2 To make the filling, combine 6 tablespoons of granulated sugar, the brown sugar, 1/2 teaspoon cinnamon, the raisins, and walnuts.

3 On a well-floured board, roll each ball of dough into a 9-inch circle. Spread the dough with 2 tablespoons apricot preserves and sprinkle with 1/2 cup of the filling. Press the filling lightly into the dough. Cut the circle into 12 equal wedges cutting the whole circle in quarters, then each quarter into thirds. Starting with the wide edge, roll up each wedge. Place the cookies, points tucked under, on a baking sheet lined with parchment paper. Chill for 30 minutes.

4 Preheat the oven to 350 degrees F.

5 Brush each cookie with the egg wash. Combine 3 tablespoons granulated sugar and 1 teaspoon cinnamon and sprinkle on the cookies. Bake for 15 to 20 minutes, until lightly browned. Remove to a wire rack and let cool.

SHRIMP ALFREDO

I found this recipe by Sarah Welch on the internet (as I now find most of my recipes). It's simple, delicious, and full of calories and protein. You may not want to study the nutritional info for too long, but it's all there if you want it.

This shrimp Alfredo pasta is made with fettuccine in a creamy parmesan sauce, topped with sauteed garlic shrimp. An easy dinner that tastes like it came from a fancy restaurant!

Course Main
Cuisine Italian
Level: Easy
Prep Time 10 minutes
Cook Time 20 minutes
Total Time 30 minutes

Servings 4
Author Sara Welch
Ingredients

- 10 ounces fettuccine pasta
- 5 tablespoons butter divided use
- 1 cup heavy cream

- 3/4 cup freshly grated parmesan cheese
- salt and pepper to taste
- 1 pound shrimp peeled and deveined, tails removed if desired
- 1 teaspoon minced garlic
- 2 tablespoons chopped parsley

Instructions

1. Cook the pasta in salted water according to package directions.
2. Melt 4 tablespoons of butter in a saucepan over medium low heat.
3. Add the cream and simmer for 4-5 minutes or until just thickened - do not boil.
4. Whisk in the parmesan cheese, stirring continuously, until cheese has melted.
5. Season the sauce with salt and pepper to taste.
6. Heat 1 tablespoon of butter in a large pan over medium high heat.
7. Add the shrimp to the pan and season with salt and pepper.
8. Cook the shrimp for 3-4 minutes, stirring occasionally, until shrimp are pink and opaque.
9. Add the garlic to the pan and cook for an additional 30 seconds.
10. Drain the pasta and toss with the Alfred sauce. Arrange the shrimp on top and sprinkle with parsley, then serve.

Nutrition

Calories: 689kcal | Carbohydrates: 52g | Protein: 41g | Fat: 35g | Saturated Fat: 26g | Cholesterol: 477mg | Sodium: 743mg | Potassium: 325mg | Fiber: 2g | Sugar: 1g | Vitamin A: 1505IU | Vitamin C: 5mg | Calcium: 454mg | Iron: 3.9mg

ABOUT THE AUTHOR

Sarah Osborne is the pen name for a physician and writer living on Cape Cod. She grew up in northern California, went to school in Ohio and lived for many years in Atlanta before moving to the Cape. She writes cozy mysteries for the same reason she reads them—to find comfort in an often troubling world.

You can find her books and more about her in the following places:

Her website (https://doctorosborne.com)
Amazon (amazon.com/Sarah-Osborne/e/B07DKN-ND23%3Fref=dbs_a_mng_rwt_scns_share)
Facebook (facebook.com/Sarah-Osborne-Mystery-Author-187314178540361)
Her monthly newsletter (sign-up at: https://landing.mailerlite.com/webforms/landing/u6y8v2)

Made in the USA
Monee, IL
19 August 2025

23699006R00166